Dennis Gabor

Innovations:
Scientific, Technological, and Social

OXFORD UNIVERSITY PRESS 1970

Oxford University Press *Ely House, London W.1*

Glasgow	Bombay
New York	Calcutta
Toronto	Madras
Melbourne	Karachi
Wellington	Lahore
	Dacca
Cape Town	
Salisbury	Kuala Lumpur
Ibadan	Singapore
Nairobi	Hong Kong
Dar es Salaam	Tokyo
Lusaka	
Addis Ababa	

Printed in Great Britain by
Spottiswoode, Ballantyne & Co. Ltd.
London and Colchester

Contents

1

Introduction

When *Homo sapiens* appeared on the Earth, more or less in his present shape, innovation began. Early Man was equipped with the same sort of brain that later could write the *Principia Philosophiae Naturalis* and the *Principia Mathematica*, but the brain was almost empty. First had to come the greatest of all inventions: language; then tools, weapons, and a primitive social organization suitable for agriculture and the domestication of animals. Then, much later, came writing, and what we now call history.

What is now called innovation still has an element in it of the instinct that drove primitive man to produce such wonderful inventions as the bow and arrow, or to devise such complicated social arrangements as totemism. Mechanical inventions and social innovations have remained indispensable but uneasy partners from prehistory to modern times. They were devised by two different kinds of human minds, and both were suppressed for long periods by the third type of man, who cared neither for technology, nor for social progress, but only for power. Regrettably, history records mostly the deeds and misdeeds of this third type of man.

We have now come to a point in the evolution of civilization when such history must have a stop. After a long and mostly tragic epoch of darkness, some three hundred years ago the empirical technology of the craftsman was joined by the systematic knowledge of nature and the method of reasoning from facts, not from fancies, which we call science. The confluence of techniques and theoretical science created applied science, which gradually became synonymous with modern technology. As we all know, this has now reached a stage at which it can destroy all civilization, at least temporarily, or create a new and happier world—if only Man is fit for such a world. The prospects for 'Post-historic Man', as painted by Roderick Seidenberg,*

* Roderick Seidenberg, *Post-historic Man*, Chapel Hill, The University of North Carolina Press, 1950.

are dim. When looking into the far future, one is torn by doubt whether Man, this fighting animal, can ever settle down to be happy. One doubts also whether such a state, in which most of our instinctive and historic values are negated and in the end even human intelligence becomes unnecessary, is worth achieving. Many of our greatest minds, such as Albert Einstein, have come to the conclusion that *Homo sapiens* is approaching his end.

But the instinct that has brought us from the naked ape in the forest to modern man tells us that we must not give up. Man has fought nature and his own kind for perhaps a hundred thousand years; now he will have to fight his own nature. Ultimately this must be the aim of any far-sighted innovator of our times and in the years to come.

At the present there is a terrifying imbalance in innovations. After many centuries in which innovation was almost imperceptible, and after a few in which all technical progress was identified with human progress, we have now reached a stage in which innovation has become *compulsive*—but only technological innovation. A large vested interest has been created, even apart from the military–industrial complex, embodied in the *avant-garde* industries and research organizations, which believes that it must 'innovate or die'. Man's landing on the Moon is the epitome of this development; a splendid triumph of applied science and of a brilliant cooperative organization, at a time when most thinking Americans are filled with grave doubts about the sanity of their society.

The time of inventions in the ordinary sense is not yet over. The scientific–technological complex that we have created will produce more, almost automatically, by its own inertia, which is very different from conservatism. It is inertia according to Newton's (really Galileo's) First Law: 'a body remains in uniform motion unless acted upon by an external force'. We can with certainty expect from it cheaper and more abundant power, an even closer communications network, and revolutionary improvements in information processing. But novelties far more important in their human and social implications can be expected from the biological sciences, which as yet have much smaller establishments. They have already upset our world equilibrium by death control; now they can upset it even further by creating enough food for many more billions of people, until the world becomes intolerably overcrowded. They will also almost certainly provide us with powerful methods of mind control, which can be used for good or evil.

Shall we be able to control the Controllers? There is no simple answer to this. The time of simple solutions is long behind us, though we are still suffering from the simplistic slogans of the nineteenth century, such as 'Free Enterprise', or 'Common Ownership of the Means of Production'. We have long ago entered the era of compromises, of the piecemeal reconcilement of human nature and of vested interests with desirable aims, national and international. Unfortunately, the struggle with these contrary forces has recently visibly shortened the 'lead times', and we can see the governments planning for months ahead, rather than for ten to twenty or more years.

If this goes on much longer, we are certain to run into a catastrophe, even without an all-out nuclear war. With the present trend, the world population will double by the year 2000—and A.D. 2000 ought not to be the end of the world. Unfortunately all our drive and optimism are bound up with continuous growth; '*growth addiction*' is the unwritten and unconfessed religion of our times. In industry and also for nations, growth has become synonymous with hope. Undoubtedly, quantitative growth will have to go on for many more years, but unless we prepare for a turning-point well before the end of the century, it may by then be too late.

History must stop, the insane quantitative growth must stop; but innovation must not stop—it must take an entirely new direction. Instead of working blindly towards things bigger and better, it must work towards improving the quality of life rather than increasing its quantity. Innovation must work towards a new harmony, a new equilibrium; otherwise it will only lead to an explosion.

There is much in Arnold Toynbee's monumental *Study of History* with which I do not agree, but I fully subscribe to his thesis that a civilization without a challenge must perish. Here is a supreme challenge for all the creative spirits of the new generation; stop the insane race towards overcrowding; stop the armaments race; stop growth addiction; form the New Man who can be at peace with himself and with his world.

The impatience of youth has recently manifested itself in the world-wide university rebellions, but I would consider this as a symptom of the *malaise* of our civilization rather than as a sign of clear awareness of its new problems. What we want is not rebels who want to smash up the institutions they do not understand, but men and women who will create new ones, patiently, but relentlessly, each in the line that he or she understands, but always with the whole world-wide vision before their eyes. Though my appeal is mostly to the elite, millions can contribute to it if they devote themselves to education instead of to the thoughtless multiplication of material goods.

In what follows, I shall first deal with material inventions and innovations, and then I shall try to outline, as best I can, the far more important social innovations. I cannot expect to succeed entirely; these are problems, not for a single man, but for a whole generation.

1.1 Science and technology, from Greek thought to the crash programme

Our Western science, which has now become world-wide, is the direct successor of Greek science, enriched by the wonderful flash of the short-lived Arab Renaissance. Indian science had little influence on it, the highly respectable body of native Chinese science even less. Its full history has become available to us only in very recent years, centuries after China had lost its lead over the West.

When Greek science was born, it was almost entirely divorced from technology, and even more from politics. It flourished for some five hundred years, during which Greek political power decayed continuously and dwindled to almost nothing. It was beyond the brilliant Greek intellects to grasp the idea that scientific power *via* technology turns ultimately into political power. Aristotle shunned social innovations, because he was convinced that any change would be for the worse. He made the famous statement that slavery would become unnecessary when the weaver's shuttles would run by themselves, but it appears that he never really thought that shuttles could run without a slave flogging them.

Science and technology remained divorced until the European Renaissance. Science not only stagnated, but was almost entirely forgotten in Europe, while technology developed under its own power in the hands of craftsmen, who made important inventions, such as the proper harnessing of horses, the iron plough, and spectacles, centuries before the Renaissance. Galileo (1564–1642), was perhaps the first man who was both a scientist and a technologist. During his lifetime the idea that science must be wedded to the 'useful arts' spread so rapidly that it was fully grasped by an outsider, the lawyer Francis Bacon, and a little later it became the programme (in modern parlance *the mystique*), of the founders of the Royal Society (1660). This was also the great epoch when the idea of progress first dawned upon thinking men.

The wedding of science and technology was not achieved in one step, but only gradually, in the course of three centuries. Even in the nineteenth century it was so imperfect that, though most of the laws of electricity and magnetism were discovered and fully formulated by Faraday and Maxwell, not a single one of the electrical machines was invented in the country of the Royal Society. Nor was it perfect in other countries. Heinrich Hertz produced, in 1887, the electro-magnetic waves that had been implicit in Maxwell's equations since 1868, but it was left to Marconi in 1896 to utilize the antenna, whose complete theory was contained in Hertz's equations. This gap of twenty to forty years between scientific discovery and technological exploitation remained typical for most of the nineteenth and early twentieth century. It was dramatically shortened only in our own times. It took only six years from Otto Hahn's discovery of nuclear fission to the first atomic bomb in 1945, and the men who led the enormous team of the first crash programme in history were them-selves scientists, moreover mostly *theoretical* physicists.

This does not mean that such long gaps cannot occur in our times, and it would be unwise for inventors to disregard scientific results that are more than a few years old. The laser was implicit in Einstein's equations of 1917, but it was not until 1958 that Townes and Schawlow laid down explicit rules for its realization. It then took only two years for T. H. Maiman to produce the first ruby laser, and for A. Javan to make the first helium–neon laser.

Such long delays occur most often when a vital component of the invention is missing. The aeroplane was visualized in essentially its modern form by the nineteenth-century inventors Cayley, String-fellow, Pénaud, and others, but it could not lift itself off the ground until the advent of the internal combustion engine. Early experiments with gas turbines failed miserably for lack of a good compressor and materials that had high strength at high temperatures. The jet engine was saved just in time by 'Nimonic' and similar alloys. Centrifuge separation of uranium-235 was contemplated early in 1940, but it became a practical possibility only about twenty-five years later with the new two-phase high-tensile materials. Holography was an aca-demic experiment in 1948, it became a success in 1963 when the laser was first applied to it.

Modern crash programmes can sometimes supply the component

inventions in 'forced marches'. When the *Polaris* submarine was conceived, four major inventions were needed for its success: the nuclear drive, accurate location under water, the solid-fuel rocket, and inertial guidance. As a fifth, one can add 'PERT', a planning scheme that made it possible for millions of parts supplied by 11 000 manufacturers to arrive in time and to fit together.

While Polaris was essentially an engineering feat, the invention of the transistor is typical of a modern development in which the border-lines between science and engineering are blurred. Shockley, Bardeen, and Brattain developed the junction transistor in ten years, starting from the wish for a solid-state amplifier, with invention always proceeding *pari passu* with the understanding of the physics of the flow of electricity in semiconductors. The scientist and the technologist have again become united, as they were in Galileo.

1.2 Quantity changes into quality

It would be highly misleading to consider the change from the individual inventor in his garret, who took his little knowledge out of old textbooks of physics and chemistry, to the modern super-team, with all grades of men, from the mathematician to the mechanic, and the compression of time from a generation or two to a few years, merely as a change in quantity. A change on such a scale means *a change in quality*, a change in the intrinsic nature, aims, and consequences of the process of invention and innovation.

There are three factors in this; the change of the time-scale, the change in the magnitude and of the social consequences of the innovation, and the change in scope or aim. Let us take them in turn.

The natural time-unit in social life is one generation. Apart from rare exceptions, people cling throughout life to the values they have made their own in childhood or as young men. Medieval man expected his grandchildren and his great-grandchildren to live as he did, with the same values. Technological change was almost imperceptible. Wars and pestilences were temporary disturbances, endured sometimes with stoic calm, sometimes with outbreaks of mass hysteria, but after these life settled down more or less in its old course. Social changes, such as the enclosures in England, had more lasting effects than the almost stagnant technology. A radical change came about only in the nineteenth century, at the time when the mechanical loom threw thousands of weavers into abject poverty, but it affected only a small fraction of the population; for the great majority it was still a stable world. It grew slowly. Even in the last century, the population in the industrial countries doubled in about a hundred years. Only in our time did the doubling time come down to about one generation, and essential changes in the mode of living to a fraction of a generation.

Second, the change in magnitude. Gunpowder, the Maxim gun, even the bomber plane with high explosive could kill only a fraction of a population; the hydrogen bomb, and perhaps also some devilish viruses bred in biological warfare establishments, could kill as good as the whole of it. The mechanical loom made a few thousand weavers jobless. Modern mechanization, rationalization, and automation could *reduce* the labour force in the United States by about a million a year, instead of having it take up the annual surplus of births over deaths. In the United Kingdom the 'saving' could be about the same, though the labour force is only about one-third that of the U.S.A., because of the considerable overmanning in British industry. Of course this will not happen, and cannot be allowed to happen so long as we do not know what to do with the millions of unemployed. It demonstrates clearly the imbalance between technological and social innovation, and the unhealthiness of a state in which most of the best brains still try to improve technology when the bottleneck has shifted long ago to the mismatch between technology and society.

Another social consequence of this imbalance must not remain unmentioned. Arthur Koestler argued in 1940, in his great novel *Darkness at Noon*, that every new invention is a threat to democracy because it makes politics unintelligible for the simple man. This may have appeared an exaggeration in 1940; it is no longer so thirty years later. The 'modern industrial state' or the 'technetronic society', as it has been variously called, is indeed above the head of the man in the street. How could the simple man decide with his vote a question, such as was put by Bertrand de Jouvenel: 'How to maintain full employment, not more than 2 per cent inflation per annum, and a good balance of international payments at a steady rate of real growth of not less than 3·5 per cent?' This is only a fair statement of the problem the government has to solve, but what political party would dare put such a question to the voters? The simple man will rather consider which party is likely to serve the interest of his class best (full employment at rising wages if he is a wage-earner, lower taxation if he is of the middle class), but is this democracy? It may still be rule *for* the people, but hardly rule *by* the people.

The third change, which is clearly of qualitative nature, has occurred in the scope and aims of inventions and innovations. Until fairly recently inventors served primary needs or archetypal human desires: grow two blades of grass where one grew before; speak to friends at a distance; travel fast; fly. Certainly modern plant breeders can not only grow twice the number of grass blades, they can even produce four times the yield of rice as before. But what is the use if the population is growing so fast that more are going hungry than before? Even if we could feed them all, what is the use if they crowd together in multi-million cities, which become explosive by their very size?

When Alexander Graham Bell invented the telephone, he could rightly feel that he had satisfied an archetypal desire of mankind. But now the telephone is with us, to the extent that we could not do without it, and men (and even more women) spend a sizeable fraction of their time on the telephone. What is there left to invent? The Picturephone, to see the speaker, not only to hear him? Yes, certainly, but then we shall have reached saturation.

Rather worse than saturation is what is likely to happen to the equally archetypal desire of man to travel fast and to fly. The motor car has crowded some cities to the extent that traffic jams are everyday experiences, and it would pay to walk—were it not for the difficulties of crossing the traffic. In the air near-misses are equally common, and collisions can only just be avoided by the most ingenious devices that men can contrive. But air traffic is now doubling every seven years, and if it goes on increasing like this it may be impossible, even with the most ingenious computer-controlled devices and with the biggest air-buses carrying 500 to 1000 passengers, to prevent either traffic jams in the airports or frequent collisions in the air. At present air transport is still by far the safest of all methods of passenger conveyance. But in 1969 the total number of passengers of 104 airlines reached 232 million, with 57 million international flights, the latter with a doubling time of less than 5 years. How long can it take before the toll of air traffic will approach the level of road deaths? Ten years or twenty? There must be a stop before that because people are not likely to tolerate in public transport what they accept in private transport.

Traffic jams and deaths on the road or in the air, to which we could add the problems of pollution of the air, the rivers, the lakes, and the sea, take us to what is probably the most important difference between the inventions of the past and of the present day. *The most important and urgent problems of the technology of today are no longer the satisfactions of primary needs or of archetypal wishes, but the reparation of the evils and damages wrought by the technology of yesterday.*

Elsewhere I have expressed this by saying that we cannot stop inventing because we are riding a tiger. Fossil fuels are threatened by exhaustion; so we must have nuclear power. Death control has upset the balance of population, so we must have the pill. Mechanization, rationalization, and automation have upset the balance of employment; what is it we must have? For the time being we have nothing better than Parkinson's Law and restrictive practices.

1.3 Introductory note to the list of inventions and innovations

When I first drew up, in a provisional form, the lists that follow,* I was greatly helped by a list of one hundred inventions and innovations compiled by Herman Kahn and Anthony J. Wiener† that had appeared a few months earlier. I am still greatly indebted to these authors. The chief difference between their list and mine is not so much in the considerable number of their items that I have omitted and of my own (chiefly social inventions) that I have added, but in the strongly *normative* approach I have adopted. They listed their items roughly in the chronological order in which the inventions are likely to arrive, and abstained from all judgements about their value. I have freely expressed my personal views of their desirability or otherwise.

I am also greatly indebted to my friend Olaf Helmer, who with various collaborators has initiated and developed the DELPHI method of forecasting technological and scientific developments, first in the RAND Corporation, now in the Institute for the Future (IFF), at the Wesleyan University, Middletown, Connecticut. The DELPHI method seeks a consensus of experts on future innovations such as, for instance, 'laboratory operation of automated language translators capable of coping with idiomatic syntactical complexities'.

---

* In two talks to the Science of Science Foundation, London, 14 and 18 February 1968.

† Herman Kahn and Anthony J. Wiener, 'The next thirty-three years. A framework for speculation', *Daedalus*, Summer 1967, pp. 705–32. This appeared a little later in book form as *The year 2000, a framework for speculation on the next thirty-three years*, Macmillan, New York, 1967, 431 pp.

The respondents are asked at which date they expect the event to occur with 10, 50, and 90 per cent subjective probability.* From this the dates are computed at which half of them put the probability at 25, 50, and 75 per cent. These are called the first quartile, the median, and the third quartile. The novelty in the DELPHI method is that the consensus is communicated to the respondents *anonymously*, so that they need not eat their words when bowing to the majority. Experience shows that in this second round the quartiles approach much closer. A third round is not usually necessary.

In the most recent publication of the Institute for the Future† there are seventy-six items. Whenever one of these clearly coincided with one of my list, I have given the IFF data in the following form.

(IFF: 1975–1985–1995, experts 1990.)

This means that the first quartile is 1975, the median 1985, the third quartile 1995, and the median of the guesses of the 'experts', a group of specialists among the respondents, is 1990. I have not given my own estimate, but it will be clear in most places that my expectations are less sanguine than those of the DELPHI respondents.

In further developments of the DELPHI method the respondents are also asked about the foreseeable social consequences of the new developments, and about their desirability. At this stage the forecasts acquire a normative aspect. Where I could compare my views with theirs, I have usually found myself on the pessimistic side.

I list in the following chapters 137 inventions and innovations: 73 of them 'hardware', 27 biological, and 37 social. In the biological section I have been leaning particularly heavily on the list of the Institute for the Future.

I am not using the term 'innovation' in the restricted sense, that is, the process that turns an invention through development, pilot manufacture, sales propaganda, etc., into a marketable product. If I had adopted this, it would have been nonsense to talk of 'scientific

* Of course nobody knows exactly what a subjective probability is. If a man puts it as 25%, this means, if it means anything, that he would accept a bet with 3 to 1 odds. But for how much? A penny or five pounds?

† Theodore J. Gordon and Robert H. Ament, *Forecasts of some technological and scientific developments and their societal consequences*, IFF R.6, September 1969. See also Olaf Helmer, Theodore J. Gordon, Selwyn Enzer, Raul de Brigard, Richard Rochberg, *Development of long-range forecasting methods for Connecticut, a summary*, IFF, R.5, September 1969, and Olaf Helmer, *Political analysis of the future*, IFF, R.1, August 1969.

innovations', and 'scientific invention' is not appropriate either in a good many cases. What I have called 'social innovations' are more usually called 'reforms'. I prefer to use innovation for all *methodical* creations of the human spirit, that is to say for all novelties that once created can be usefully and repeatedly applied.*

Innovation in the restricted sense is a powerful built-in agency in our social system: 'Innovate or die!' But the most successful innovations are not those that I would consider as triumphs of the creative spirit. They are the improved *substitutes* for products that already have a market; ballpoint pens for fountain pens, aerosol cans for tubes. These are trivial in a way, but immensely important for our economic system:

Capitalism, then, is by nature a form or method of economic change and not only never is but never can be stationary. And this evolutionary character of the capitalist process is not merely due to the fact that economic life goes on in a social and natural environment which changes and by its change alters the data of economic action; this fact is important and these changes (wars, revolutions, and so on) often condition industrial change, but they are not its prime movers. Nor is this evolutionary character due to a quasi-automatic increase in population and capital or to the vagaries of monetary systems of which exactly the same thing holds true. The fundamental impulse that sets and keeps the capitalist engine in motion comes from the new consumer's goods, the new methods of production or transportation, the new markets, the new forms of industrial organization that capitalist enterprise creates.

This was written not by Karl Marx, but by Joseph Schumpeter, who was very far from being a socialist.† In this passage, and in many others, he makes it clear that if there is no innovation not only will the firm that neglects it die, but so will capitalism itself.

I can only repeat what I have said before, that capitalism may stop with the end of the 'ballpoint pens for fountain pens' type of innovation, but evolutionary innovations must not stop.

---

* This excludes artistic creations. A symphony by Beethoven can be repeated but not applied. On the other hand novelties in orchestration can be applied and can be counted as innovations.

† Joseph A. Schumpeter, *Capitalism, Socialism and Democracy*, 3rd ed. Harper Brothers, New York, 1950.

'Hardware' inventions and innovations

2

These are inventions and improvements in the 'classical' line, those that are patentable under British or U.S. law. There is no hard-and-fast division between inventions and improvements. In fact, the list below consists almost exclusively of improvements, not first fulfilments of archetypal wishes such as the first telephone or the first flying machine was. This is due not merely to the lack of imagination of the author and his sources, but mainly to the exhaustion of primitive desires by past inventions. There remain some primitive wishes such as ESP (Extra-sensory perception, for direct communication from mind to mind), telekinetics (moving objects by wishing), the time machine, antigravitics, and super-photonic speed (breaking through the light velocity barrier), but on all present evidence these will remain in science fiction—until even SF becomes tired of them.

When talking of inventions that may (or may not) mature in the future, I had to give a rough account of their present state. This I have tried to do with the assistance of experts, whom I mention in the appropriate places. But what I deeply regret is that I could not give due credit to the inventors. In the case of inventions that are still *in statu nascendi*, many in great firms or in government establishments, it is difficult and often invidious to give credit to an individual. This will be a job, and often a difficult one, for historians of the future. I regret this all the more as I am an inventor myself.

I have quoted few books, and hardly any journals. References to these would have soon become obsolete. Most of the subjects are in a state of such rapid flux that the reader will find new developments every week in the trade journals, and even in the daily newspapers.

2.1 Materials

1. *High-strength materials*

These are composite, two-phase materials consisting of very fine fibres (carbon, beryllium, quartz, and others have been tried) in a matrix, usually a plastic. Stiff, strong carbon fibres with exceptionally high tensile strengths and elastic moduli have been discovered and developed, and are now being produced from polyacrilo-nitrile by the Royal Aircraft Establishment (Farnborough) with help from the United Kingdom Atomic Energy Authority (Harwell) and Rolls-Royce (Derby). These fibres are used to reinforce plastics. Rolls-Royce are fabricating the large compressor blades of their subsonic transport RB 211 engine from such material. The trade name is 'Hyfil'.

There are many important potential applications, especially in the aircraft industry, but their rate of introduction depends upon the development of low-cost large production facilities. A Parliamentary Committee has recently recommended the setting-up of further carbon-fibre production plants, to reduce the cost of the material to less than 10 shillings per pound weight.*

One of the most important, though the least desirable, applications of this material may be in centrifugal separators for uranium-235, in which even the strongest metals have given entirely unsatisfactory yields. At the time of writing two centrifuge separator plants are projected, one at Capenhurst, England, the other at Almelo, Holland, both to operate in 1972. News has just arrived about a Chinese atomic centrifuge plant. Uranium-235 is the detonator of hydrogen bombs, and at present it can be produced only in enormous thermal diffusion plants. If we are unlucky enough, these wonderful new materials, prominent among British export assets, might produce the 'poor man's hydrogen bomb'.

(IFF: whisker-reinforced composites at competitive prices, 1975–1980–1990, experts 1975.)

2. *Heat-proof, high-strength plastics; high-performance fabrics*

In addition to a large growth of general-purpose plastics (e.g. polythene, polystyrene, and PVC), modified forms of these, together with the so-called 'engineering plastics' like nylons, acetals, and their successors, will be used increasingly in car and boat bodies, domestic durables, office equipment, building products, etc. These products will be lighter and impervious to corrosion, and will require fewer man-hours to produce than their equivalents in metal.

Plastics like the polyimides and fluorocarbons will give special properties like high temperature resistance, inertness, and low

* I am indebted for this information to S. L. Bragg, Chief Scientist, Rolls-Royce, and K. D. B. Johnson, Head, Process Technology Division, UKAEA, Harwell.

friction without lubricants. Polymers in fibre form will be engineered to provide clothing and domestic fabrics, eliminating tedious traditional weaving processes. 'Super-fabrics' as developed for space exploration will have increasing industrial use.*

Cross-linking polymers by nuclear radiation to produce plastics resistant to extreme heat has so far proved rather disappointing. Nevertheless, this may be a line worth exploring, because there are indications that irradiating organic substances can confer extraordinarily high dielectric strengths, and perhaps lead to condensers as 'light accumulators'. The great Soviet physicist A. Joffe forty years ago considered these as realizable in the not too distant future.

3. *Precision casting, forging, and moulding*

Not only aluminium and brass but iron and steel can be cast with such precision that the cutting operations can be reduced to a minimum. Explosive moulding, cladding, and hydrostatic extrusion have produced results with the most difficult metals which not long ago would have appeared unbelievable; this is an actively developing field, promising complicated metal components almost as light and almost as quickly produced as plastics.

4. *Ultra-high-pressure processes combined with high transient temperatures*

These processes are successful in producing diamonds and borazon. So far, few materials have been found that undergo an irreversible change by ultra-high pressure alone, but combination with high transient temperatures may well produce a range of 'frozen-in' compounds of industrial interest.

Quartz has two interesting compressed forms, coesite and stichovite. Both were first found in meteorites, the first by Loring Coes, the second by S. M. Stichov of the Moscow State University. In stichovite, which was reproduced with a pressure of 160 000 atmospheres at 1200°C, each silicon atom is surrounded by five equidistant oxygen atoms instead of four.

5. *Substitutes for rare metals and methods for their recovery*

So far, the most successful substitutes for rare metals, such as tin, are plastics. There is no imminent danger of exhausting the rare metals, such as copper, lead, tin, or zinc, but shortage can be expected before the end of the century. By that time, copper may be mined on a

---

*I am indebted for this information to A. E. Willbourn, Director, Plastics Division, ICI, Welwyn Garden City.

large scale from the ocean floor, but the use of lead and tin, which are now wasted in an irresponsible way, ought to be restricted and methods be made available for their recovery. Cheap nuclear power has now made aluminium a relatively cheap metal, and it is also easily recoverable.

6. *Water*

Adam Smith (1776) mentioned water as the paradigm of a good that is enormously useful and yet has no commercial value. This situation has now radically changed. The Isle of Malta must be supplied with water by ships. The entire existence of some countries like Egypt is bound up with the water supply. In the industrialized countries, the pollution of the rivers has created critical situations; for instance, in Germany, where in 1969 the waste products of a single factory killed all the fish in the Rhine. Powerful methods for purifying water of the normal refuse of cities are available, based on digestion by aerobic and anaerobic bacteria, which convert the waste into a fertilizer (activated sludge); but new waste products, such as some detergents, create new problems.

The desalination of sea-water, of vital importance, not only for desert countries, but also for southern California, is now almost entirely a problem of cheap power and cheap money. The power for the two principal methods, distillation and freezing, can be obtained cheaply enough from nuclear plants, but the installations are so expensive that they can bear only an interest rate of 2 per cent per annum. So far, it is economic only in exceptional places (such as Kuwait), but with the increase of population, almost any price will forcibly become 'economic'. Its use for agriculture in arid zones will start to pay earlier if the water is not just left to evaporate but is kept below the surface, covered with pebbles and with a thin layer of an evaporation-intercepting substance.

By the end of the century, the problem of pure and abundant water may well exceed in importance all other material problems.

(IFF, 20 cents/1000 gallons: 1973–1980–1985, experts 1980.)

7. *Pure air*

Most of the big cities have dealt successfully with the soot problem by establishing smokeless zones and forcing industry to put a stop to factory chimneys belching black smoke, but they have not yet dealt with air pollution by motor cars. Pure air is a problem different from the others in this section because the question is not one of manufacturing pure air, or of purifying it, but of stopping pollution. We will deal with it under the section on motor cars.

Another atmospheric problem is being hotly disputed at the present time; the question of the CO_2 content of the atmosphere. There is some evidence, though not yet quite convincing, that by the end of the century the CO_2 content might increase sufficiently to produce an appreciable change in the climate, because combustion is producing more and more, and the somewhat reduced area of green vegetation can no longer eliminate it. To the end of the century, this is likely to be beneficial, but if the process continues and the polar caps begin to melt, it may create a very difficult international problem. Perhaps it will be dealt with by covering a part of the oceans with green vegetation? (Even at present the microflora of the oceans produces about 70 per cent of the oxygen; only 30 per cent is produced by land vegetation.)

2.2 Power

Industrial civilization lives on power. John Kettle's estimate that the cost of power is about to overtake the food bill may be a little premature, but is likely to come true before A.D. 2000. Power consumption *per capita* is a reasonable measure of the standard of living (though hardly of the 'quality of life'). At present North America, with 6·5 per cent of the world's population, consumes about 35 per cent of the world's energy. Its share will be probably much lower by the year 2000, because the consumption *per capita* increases in North America only by 2 per cent per annum while in the less industrialized half of the world the increase is more than 4 per cent per annum.

The total energy consumption is usually measured in millions of metric-ton coal equivalents, MMTCE. According to Shell International's *Information Handbook* 1969–70 the world consumption in 1968 was 4371 MMTCE; according to the UN *World Energy Supplies* about 6000 MMTCE. There is a rather wide gap between these estimates, but all agree more or less that electricity consumption increases at 8 per cent per annum. According to Sir Harold Hartley's estimate (*New Scientist*, 13 November 1969) the figures for A.D. 2000 are likely to be

Solid	Liquid	Gas	Nuclear and Hydro	Total	
4500	9500	8000	8000	30 000	MMTCE

In spite of these rather terrifying figures there is no imminent danger of energy shortage.

8. Fission power

The greatest success story of modern engineering. In just twenty years of development, nuclear electric plants have beaten fossil fuel plants not only in cost per kilowatt hour, but also in plant cost per kilowatt. The latest plants in the United States can supply a kilowatt hour off the plant for 0·3 cent = 0·3d. In fifty years, the price of electricity has dropped to rather less than a fiftieth in real value.

At the present time, the most successful reactors are the light-water boiler-type reactors whose installation cost is below that of the slightly more efficient advanced gas-cooled reactors. At the present low cost of uranium (around $15/kg in the form of uranium oxide), they are likely to remain dominant until the end of the century, when they will be perhaps replaced by breeder reactors which are more economical in the use of fuel, as they convert a fraction of uranium-238 into the fissionable component, uranium-235; but this is somewhat doubtful in view of the abundance of low-grade uranium ores, to be mentioned in Section 9.

Various authorities are in fair agreement that in 1985 20–25 per cent of the world's electric power will be generated by nuclear stations, and in 2000 40–50 per cent.

The success of fission power is no unmixed blessing, which is not surprising, seeing that fission power is itself a spin-off of the nuclear bomb. There is general agreement, even between the U.S.A. and the U.S.S.R., that the proliferation of nuclear powers is one of the greatest dangers to peace. But every nuclear plant exported to a nation in the stage of adolescent nationalism is (with some additions) a potential plutonium factory, and though plutonium bombs are midgets compared with hydrogen bombs, they might be quite sufficient to release a world conflagration.

9. Exploitation of low-grade uranium ores and of the oceans for uranium

Experiments in Ranstadt (Sweden) have shown that granite with 0·03 per cent uranium can be exploited at about twice the world price for uranium. R. Spence and his collaborators at UKAEA (Harwell) have produced convincing arguments to show that the oceans, which contain about 0·3 mg of uranium per cubic metre, could be exploited for not very much more. The oceans are a virtually inexhaustible source of uranium; thousands of tons of uranium are washed into them every year by the rivers. They are capable of supplying energy for millions of years for mankind, however extravagant it may be.

10. *Nuclear reactors that cannot be used for the manufacture of atomic weapons*

I mention this as a highly desirable innovation, of which, unfortunately, there is not much chance. All nuclear reactors produce plutonium as a by-product, even the breeders. At the end of the Second World War, when the prospect of nuclear power first emerged, the atomic scientists proposed a 'contamination' of fuel rods which would have made the plutonium useless for atomic bombs, as it stopped the fast neutron reaction; but this, like any other suggestion that would have involved inspection, was of course, rejected. Though a nuclear reactor that is useless for producing plutonium would be probably also rejected by nations determined to possess atomic weapons, at least it would stop or slow down the stockpiling of plutonium by those who already possess atomic bombs. The present-day reactors almost *force* the military to use the by-product and to stockpile it, and the sale of plutonium is still a non-negligible part of the profit of nuclear power plants.

11. *Controlled fusion*

After the success story of fission, the sad tale of fusion power. In the early 1950s it was a general belief that only 'burning the sea', that is to say producing power by the fusion of the heavy hydrogen contained in the oceans, could produce enough power for industrial civilization for a million years. Great teams of outstanding scientists attacked the problem in the U.S.A., in Britain, and in the U.S.S.R., armed with computers and with some of the biggest and most ingenious machines ever built. The result was one of the greatest disappointments in the history of science and of engineering. The confident predictions of the scientists were proved wrong by sometimes as much as a factor of a million, and all the surprises were against the inventor. Nature presented science, for the first time on a large scale, with the phenomenon of *collective interactions*. In order to achieve fusion, a plasma of deuterium and tritium had to be contained in a 'magnetic bottle' and heated up to a temperature of many million degrees. The magnetic bottle was so cleverly contrived that not a single ion could have escaped; but they escaped in bunches, by collective conspiracy.

After about twenty years of frustrated efforts, the funds for fusion research have been gradually reduced in the U.S.A. and in the U.K. It appears, however, that in the U.S.S.R. fusion is treated as a prestige project with high priority, because they have disclosed the news of their TOKAMAK 3; an extraordinary *tour de force*. It has considerably outstripped British and American efforts by realizing a temperature of six million degrees, a carrier density of several $10^{12}/cm^3$, and a containment time of 20 milliseconds. This, however, is still many orders of magnitude below what is required of a self-supporting fusion device, which requires 100 million degrees, a density of at least $10^{15}/cm^3$ and a containment time of about a second. It is alleged that the Russians are planning a 'TOKAMAK 10', but this would be a daring and costly engineering feat, requiring a magnetic field, in a very large volume, of 50 000 gauss and an input power to start it up of 200 megawatts. It appears somewhat doubtful whether prestige would be worth this effort, because fusion power is now unnecessary. The rocks and the oceans present an inexhaustible source of fissionable material.

It is reported that by exploding lithium deuteride icicles with laser beams Soviet scientists have come within three orders of magnitude of a self-sustaining fusion reaction. This must be judged for the time being with some reserve.

There is some talk that fusion of the hydrogen atoms picked up in interplanetary space by the galactic vehicles of some future century may provide their driving power, but for the present, this can be classed as science fiction.

(IFF: Laboratory demonstration of controlled fusion, 1980–1985–1990, experts 1980.)

12. *Hydromagnetic power*

Also called 'magneto-hydrodynamic power' (MHD). Another great disappointment. In the early 1950s, A. Kantrowitz pointed out that thermal power plants are very wasteful, because the temperature of the flame gases is very much higher than is necessary for raising steam. One cannot make good use of this high temperature in rotating machinery (gas turbines are notoriously inefficient), but it may be possible in a device in which there are no solid moving parts. It was first shown by Faraday that if a conducting fluid is moving in a channel in a magnetic field at right angles to it, an electromotive force is generated at right angles to both. If this is loaded with an outer circuit, the kinetic energy of the fluid is directly converted into electric power. Flame gases are not sufficiently conductive, but they can be made so by 'seeding' them with salts of the alkali metals. In such an MHD rig, the thermal energy of the flame gases can be extracted and converted into electric power until they become too cold to be conducting, even with seeding. Unfortunately, this occurs at a rather high temperature, around 1800°C; hence, the MHD rig must be operated with very hot flame gases. This was one reason for the failure of the 'open-cycle' system. No ceramic material could be found that could withstand flame gases around 2500°C sweeping along it with supersonic speed. The electrodes themselves, which were made of graphite or carbon, burned off fast.

Moreover, as the heat losses to the walls were enormous, such a rig could offer good efficiencies only in large sizes, with thermal powers of the order of hundreds of megawatts. This, in turn, led to a difficulty of much the same nature as that which frustrated the fusion project; in a wide channel, the gases refused to flow smoothly and produced violent instabilities.

The 'open-cycle' rig was proposed as a 'topping' device for fossil fuel plants. It was expected to raise the overall efficiency from about 40 per cent to 52 per cent. By the mid-1960s it became clear that even if the system had functioned as expected and with a reasonable life, it could not have competed commercially with nuclear plants. This led to the abandonment of the large-scale development in the U.S.A. and in the U.K.

Another type, the 'closed-cycle' rig, is free from some of the difficulties. In this system, an inert gas (helium) is permanently ionized by caesium vapour which circulates in it, without ever being used up. There are no chemical reactions and the cycle starts at about the temperature at which the open-cycle system has to stop. An experimental system, designed by B. C. Lindley, has been operating for several years in Newcastle and is still in working order.

Though large-scale development has stopped in the U.S.A. and in the U.K., there is still work going on. The Americans are continuing to find special applications and there is a recommendation by the Office of Science and Technology to support commercial MHD power with $2 000 000 per year for three years if industry will contribute an equal amount. The U.S.S.R. (Moscow High-temperature Institute) is building a large pilot plant (25 MW electrical) and there are limited programmes in Italy, Germany, Poland, and Japan.*

13. *Fuel cells*

Another sad tale of high hopes disappointed, though with a modicum of success. Soon after the discovery of electrolysis, the idea emerged: 'Could one not realize electrolysis in reverse?' The electric current separates water into hydrogen and oxygen. Could one not in some way burn hydrogen in oxygen so as to produce water and electric current? This would make it possible to convert chemical energy directly into electrical power without first having to degrade the chemical energy into heat.

The old idea was taken up more than a hundred years later by Francis Bacon (Cambridge) with great energy, and his efforts, assisted in Britain by the NRDC (National Research Development Corporation) and later on an immense scale in the U.S.A. by NASA, led to the first practical fuel cell—but which unfortunately remained also the last one. The Bacon cell burns hydrogen in oxygen in a pressure vessel with porous nickel electrodes; the electrolyte is caustic potash. Not only must the pressure be high, but also the temperature; the Bacon cell operates at 250°C, and the gases must be very pure. It is not, therefore, useful for commercial applications, but it was a great success in the Apollo project where it supplied the electrical energy. The cost of its development in the U.S.A. is estimated at $100 000 000.

The basic principle of generating electricity with a fuel cell is so simple that it was too readily assumed that one could operate such a cell with a cheap hydrocarbon fuel, at reasonably low temperatures, with a suitable catalyst. In fact several years of intensive work in the United States to the tune of more than $50 000 000, and several million dollars outside the United States, have failed to produce a catalyst which could have made an electric motor car driven by a fuel cell possible. No catalyst lived long enough; they were all easily poisoned, not only by CO, but even by small traces of CO_2 in the air breathed by the fuel cell. But the very fact that there were so many unsuspected problems on the fringes of the sciences involved, which emerged during this research, makes one think that perhaps science

* For the information in the last two paragraphs I am indebted to Dr. B. C. Lindley, Director, Electrical Research Association, Leatherhead.

will yet break through the remaining barriers some day, and that it is too early to abandon all hope.*

If all these efforts fail, one can get some cold comfort from the fact that in the meantime 'the grapes have proved sour'. The most attractive feature of the electric motor car was the ideal driving characteristic of the d.c. motor. But careful studies have shown that the power-to-weight ratio of these motors was far too poor, and that one would have to go to a.c. motors, with inverters, such as thyristors, to convert d.c. into a.c. with variable frequency. This will hardly be the car which 'a child can drive safely'.

14. *Thermoelectric power*

Another story of frustration. The Seebeck effect, the phenomenon whereby two metal strips joined together at two ends and one junction heated causes an electric current to start circulating, had been known since the early nineteenth century. The best pair, bismuth–antimony, at the best temperatures, operated with an efficiency of about 0·8 per cent, and nothing better could be achieved with metals.

In 1929, the great Soviet physicist A. Joffe had the intuition that much better results could be achieved with semiconductors, and he founded an institute for research on thermoelectric power. This led to the discovery of bismuth telluride (p and n) with an efficiency of 4·2 per cent. This, in turn, induced many physicists in Europe and in the United States, in particular, Clarence Zener, the research director of Westinghouse, to a considerable effort directed toward an electric motor car. The target was an efficiency of 12 per cent, which is that of an internal combustion motor car in a city. By its superior driving qualities and the absence of all gear, the small electric motor car might have well become competitive if this target had been reached. Sad to relate, after many years of research ranging through almost all binary semiconductors and a good many ternary ones, this target was never reached. The only satisfaction of the dedicated researchers was the development of some small portable power sources and auxiliary power sources for space vehicles; also some small cooling devices in which the inverse (Peltier) effect is used to cool a junction by an electric current driven through it in a certain direction.

* I am indebted for information on the present situation of fuel cells, to J. J. Crawley, of the NRDC, and to J. C. H. Hart, General Manager, Energy Conversion Limited, Basingstoke, Hants, England, the principal managers of its development in Britain.

15. *Thermionic power*

A story very similar to the last one. The great snag in thermo-electric devices was that in parallel to the heat carried by electrons (which constitutes an electric current) there is a much greater fraction carried by ordinary (or 'lattice') conduction without a current.

But what if we made the electrons go through a vacuum or gas space? In a vacuum, there is no heat conduction, and in a diluted gas only very little. There is a radiation loss, but there is no law of physics to forbid that the electrodes shall be almost perfect mirrors which radiate very little; so all we have to do is to take a hot cathode, put it at a very small distance from a collector electrode, so that the space charge shall not limit the current too much, or make the distance a little larger, and ionize the gas to suppress the space charge, and we have a direct 'thermionic' converter of heat into electricity.

In fact, after only a few years' work, the thermionic converter with caesium vapour in its gas space started off where the thermoelectric converter had to stop, and easily topped 12 per cent in efficiency. In 1966, after some more years of work, the research team of the General Electric Company led by Volney Wilson presented evidence of 20 per cent efficiency; yet the thermionic converter never became a success. It was a precision device which could be made only in small units. In the GE converter with 20 per cent efficiency, the distance between the emitter and the collector was only 0·1 mm; and the emitter was at white heat. For a power of the order of 10 kilowatts, about a square metre would have been needed, and what engineer would dare to guarantee that the hot, squirming emitter would not buckle by 0·1 mm at one point and short circuit the device? So the caesium converter was produced industrially only in small units as an auxiliary power source for satellites; and even these were not used because preference was given to barrier photocells, in spite of their lower efficiency, which did not require a mirror system for concentrating the sunlight on the emitter. Converters in which the emitter was heated by flame gases instead of by radiation were even less successful, because in spite of years of determined effort, no material could be found which did not leak after 2000 hours at a temperature of 1500°C.

This story of frustration is one of which the author has had personal experience. It appears likely that large thermionic power units, if they will be used at all, will be used only in space vehicles with nuclear reactors as primary power, because in space there are no vacuum difficulties. If this project (now held up because of more urgent problems) is taken up again, the work which I, with my collaborators M. J. Albert and M. A. Atta, have carried out (1964–6) at Imperial College (with the assistance of the NRDC and the Consolidated Controls Corporation, Bethel, Connecticut, U.S.A.) may perhaps be helpful in *one* point. We have developed new emitter materials which do not require caesium atoms to condense on them in order to emit 50–100 A/cm². Consequently, they can operate with very small caesium pressures, just enough to break down the space charge; not of the order of one torr, as in the standard caesium converters, but 0·01 torr; hence, the distance between the emitter and the collector can be increased to one mm, which is a safe distance, even with a squirming emitter. But if I am mistaken in my hopes, I would not take it too tragically. An inventor must be happy if one out of ten of his schemes succeeds.

I have listed five unsuccessful lines of research. Projects 12–15 all started with the highest hopes, which in the best case led only to some special applications. We must conclude that for a long time, nuclear boilers with rotating electrical machinery will dominate large-scale power production, and the internal combustion engine will remain the main drive of motor cars. This raises the question: 'For how long and at what price will fuel for motor transport be available?'

16. *Commercial extraction of oil from shale*

According to some recent estimates, shale may cover the oil supply of the world, at extrapolated growth, for 150 years. At the present state of the art, it would become profitable at three times the present oil prices. This is not likely to be reached by the end of the century.

17. *Artificial fuels (such as natural gas into oil)*

The technology for the synthesis of methane into long chain hydrocarbons is available (high pressures and temperatures with certain catalysts) but the costs are excessive, even with the cheapest power now available. It is doubtful whether such processes will become profitable before the oil shales are exhausted, because the earth gas supplies are likely to be exhausted first. Ultimately, when all natural supplies are exhausted, even the synthesis from the hydrogen of water and from the carbon of the CO_2 in the air will not be considered too costly to make aeroplanes fly.

2.3 Chemistry

Chemical and physico-chemical methods play a part in all innovations concerning materials, which are mentioned in other sections, but their importance is such that they deserve a more comprehensive survey, which I owe to my colleague, Professor R. M. Barrer, F.R.S.

18. *Physico-chemical techniques of separation*

Over a period of about forty years a number of new and sensitive separation techniques have appeared, most of which still continue to grow. They are listed below in chronological order.

(i) The electrolytic method of enriching water in deuterium (e.g. Lewis and McDonald).

(ii) The chemical exchange-counterflow method of separating stable isotopes, S,N,O, etc. (e.g. Urey).

(iii) The thermal diffusion–convection method of separating stable isotopes (e.g. Clusius and Dickel).

(iv) The post-war development of gas–liquid, gel-permeation, and other forms of chromatographic separation of very great selectivity (e.g. Martin and Synge).

(v) The development of molecule sieving, using porous crystals as sorbent-filters (e.g. Barrer).

(vi) The development of reverse osmosis as a separation method (e.g. Loeb and Sourirajan).

This list of inventions is by no means a full one. Even those techniques which have not grown beyond the laboratory stage can have profound impacts upon industrial practice, because they are making industrial research so much more penetrating and precise.

19. *Novel catalysts*

This too is a field in which chemists have been, and are continuing to be pre-eminently inventive. Three examples may be given.

(i) Alumina–silica cracking catalysts first introduced in the late 1920s and early 1930s for cracking and reforming petroleum crudes, to make them into more appropriate fuels.

(ii) The post-1960 molecular-sieve-based superactive cracking, reforming and hydro-forming catalysts, which have produced a major breakthrough, particularly in American petroleum technology.

(iii) The stereo-specific post-war catalysts for synthesizing stereo-regular polymers (Ziegler and Natta). Such polymers have valuable properties, much less well developed in the stereo-irregular form.

20. *Recent chemical discoveries in the inorganic field*

(i) The discovery of the chemical reactivity of the heavier so-called inert gases, krypton and xenon, which has illustrated so well the danger inherent in the label 'inert'.

(ii) The discovery of various organo-metallic compounds which can fix molecular nitrogen at room temperatures. This fixed nitrogen was reported as being too firmly held for its sub-sequent transformation to useful products such as ammonia or hydrazine. Recently, however, van Tamelen has found a fixation of molecular nitrogen which does lend itself to the subsequent ready formation of ammonia. The possibility of chemical methods for utilizing readily this most abundant element has great interest.

(iii) Other organo-metallic compounds have been found to be capable of activating molecular hydrogen in homogeneous systems, and thus making them able to participate in hydro-genation processes. Certain organo-metallic compounds are considered as model substances for enzyme action because of their peculiar properties as catalysts, such as are illustrated above.

21. *Biological chemistry*

The organic chemist has joined with biologists and crystallo-graphers in making biological chemistry and molecular biology one of the most actively growing and exciting frontiers of science. There is also little need to emphasize the role of organic chemistry in medicine and agriculture. The inventiveness is very great, and has called forth criticism as well as praise. The organic chemist also joins with the polymer scientist in the molecular design of polymers possessing novel properties.

22. *Control, testing, quality maintenance*

There is a massive and continuing role of chemistry which is not spectacular, but which is every bit as vital as are eye-catching discoveries. This is in control, testing, and maintaining of quality. It is equally present in the constant drive to improve materials little by little in all their diverse uses by 'tailoring' them in chemical ways designed to form more appropriate chemical structures.

23. *Petroleum technology*

This is a particularly impressive example of the part which the chemist can play in solving not only technological but sociological problems, such as air and water pollution. Detergents made from branched-chain hydrocarbons are not destroyed by bacteria, whereas those made from *n*-paraffins are. In 1968 the petroleum industry separated over 500 000 tons of *n*-alkanes by means of zeolite sieves, which have been mainly used for detergent manufacture.

If this process could be introduced, on an even grander scale, into the oil industry, it would have a great impact on the problem of air pollution. Straight-chain hydrocarbons of intermediate lengths form excellent diesel fuels, while shorter, branched-chain paraffins have good anti-knock properties in motor fuels, and would give high-octane performance without lead additives. Moreover long-chain, non-branching paraffins are very suitable for the feeding of protein-producing micro-organisms. It is of course a great step from the half-million ton detergent industry to the multi-million-ton oil refineries.

2.4 Transport

The motor car with the internal combustion engine has not yet reached its apogee. 55 000 road deaths per year in the United States, 1·7 million since the beginning of motor transport, with hundreds of thousands of cripples every year, do not appear too high a price to pay for *individual* transport. But the usefulness of the motor car in terms of time saved is declining so rapidly that the apogee cannot be very far away, and public transport is bound to gain in relative importance.

24. *Electric motor cars*

The pollution of air by motor cars with internal combustion

engines (unburned hydrocarbons and nitrose products) has evoked a strong interest in electric cars. Unfortunately, this is also one of those fields in which nature is against the inventor. The failure of the hydrocarbon-burning fuel cell has already been reported. Attempts toward primary cells with high power-to-weight ratios, such as the sodium–sulphur battery (Ford), or the lithium–chlorine battery (General Motors), looked *a priori* rather doubtful, because of their highly dangerous constituents, and it appears that work on them is now being abandoned.

The rechargeable zinc–oxygen battery (Joseph Lucas Ltd.)*, which can be considered as a cross between a fuel cell (because it burns atmospheric oxygen) and the accumulator (because it is rechargeable by re-plating the zinc from the solution), still appears to have better chances, but this, too, has proved a difficult development. Catalysts must be employed at the anode in recharging, and these are easily destroyed. Dendrites grow on the zinc electrode, which short-circuit the battery. It would be too early, though, to say that these problems cannot be solved. †

At the present, the best prospect is still the lead–acid battery in its modern, much lightened form, with thin lead plates, which is sufficient to drive a car for about forty miles. This may, therefore, offer an acceptable solution for urban traffic if the spent battery can be exchanged for a freshly charged one at any filling station.

25. *Motor cars with greatly reduced air pollution*

There appear to be two ways of reducing air pollution by internal combustion engines: after-burners and combustion control. It appears that after-burners have failed because no material of reasonable price could stand up against the high temperatures and the chemical reactions that accompany total combustion. On the other hand, computer-controlled injection, so adjusted as to ensure almost complete combustion at the actual running condition of the car, appears to be a problem that has been solved. It awaits only legislation enforcing its use, as car owners are not likely to bear the unavoidably increased prices of their own accord.

(IFF: 1980–1990, experts 1980.)

---

* I am indebted for information on the zinc–oxygen battery to Mr. W. J. Arrol, Director of Research, Joseph Lucas Ltd., Monkspath, Warwickshire, England.
† In July 1970 SONY Corp. announced an experimental car with a 3 kW rechargeable zinc–oxygen battery, but it occupies half the volume of the small car.

26. *High-speed rail transport on wheels*

From town centre to town centre. The model for this is the TOKAIDO (Tokyo–Kyoto) railway, with speeds up to 120 miles per hour (imperceptible for the traveller), and air-conditioned coaches with adjustable aeroplane-type seats. This type of transport alone would be sufficient progress to the year 2000 if it were introduced wherever needed, although the fares are at present rather high.

Inside towns. Many big cities still have no underground railways, or have none on a sufficient scale. New lines, like the London Victoria Line, are unprofitable in themselves, but profitable for the community by their social benefit. There exists no better method for avoiding traffic jams in cities.

27. *Super-speed wheelless transport on rails*

Vehicles with linear induction motors float above the rail or rails, and may achieve speeds of 300 miles per hour or even more. In Britain, Professor Eric Laithwaite's prototype vehicle is expected to run in Cambridgeshire in 1970. In the United States, the Garrett Corporation, under contract to the Department of Transportation, has constructed a demonstration test vehicle with 3750 pounds thrust, which is expected to reach 250 miles per hour. An experimental monorail using linear motors is already in operation in the U.S.S.R., presumably at somewhat lesser speeds. Large-scale projects with Hovercraft vehicles are on the way also in France and in Japan. The cost of wheelless transport must be expected to exceed that of railways, but it can be expected to be competitive with inter-town air travel.*

28. *Roadless land transport*

Hovercraft on land may become important for less developed, or not highly populated, countries, especially where there are streams or lakes to be crossed. At present, roadless land transport is at a disadvantage against air transport.

29. *High-speed sea transport*

Hovercraft, hydrofoils, and submarines are in competition with aircraft for fast transport, and with container ships for cheap

---

* For information, I am indebted to Professor Eric Laithwaite, Imperial College, London.

transport. It is still doubtful whether an ecological niche can be found for them, apart from a few special applications.

(IFF: surface-effect ships widespread, 1990–2015, experts 2000.)

30. *Container ships*

Their future is not in doubt. Here is a typical case of a simple mechanical innovation with enormous consequences. 90 per cent of the dock labour can be saved, and 70 per cent of the crews, owing to rapid turn-round. They are spreading rapidly in spite of resistance from trade unions, and are likely to be further perfected.

31. *VTOL* (*vertical take-off and landing aircraft*)*

These have been under serious discussion for a decade but are still in the stage of feasibility studies. No design has been started nor metal cut. For non-military craft the current talk is of an aeroplane with a range of about 500 miles at 400–500 miles per hour, carrying 100–150 people, and weighing about 100 000 pounds. Speed would be about three times that of helicopters, with much greater range and carrying capacity.

'STOL' (short take-off aircraft) have been operated for some time by Eastern Airlines, but have not proved viable. There may well be an ecological niche both for VTOL and for STOL once the air traffic density rises to two or more times the present level.

(IFF: VTOL/STOL, 1980–1990, experts 1990.)

32. *Supersonic air transport*

A characteristic example of a modern invention that had to be developed in the face of enormous difficulties, without any assured commercial future, by the fear of competition, and by the necessity of keeping the technical staff of aircraft companies employed on a technically exciting project. The Anglo-French Concorde and its counterpart in the U.S.S.R. have now completed supersonic test flights. Their commercial future is still uncertain. The more ambitious SST of the Boeing Company, with variable geometry, had to be abandoned, and will now be replaced by a 'Delta' model. The question of the sonic boom has not yet been satisfactorily cleared.

33. *Air buses*

Giant planes carrying 500 passengers in the first generation, perhaps 1000 in the second, are not an attractive proposition, either

---

* For information regarding VTOL I am indebted to Dr. S. G. Hooker, F.R.S., Technical Director of Rolls-Royce Ltd., Bristol Aircraft Division.

for pilots or for airports, but they are unavoidably needed, or else the airports will not be able to deal with the passenger traffic which is now doubling every seven years. The Boeing 747 is the first of these, and its commercial future appears assured. In the first years, considerable difficulties can be expected, as the airports have not yet decided to deal with these aeroplanes by methods radically different from those which were (barely) sufficient for aircraft with 200 seats. Ultimately, the passengers will have to be taken to and from these giant planes with fast trains shuttling between the airports and the air terminals (which will have to become more numerous than they are today) and the luggage will have to be containerized to be dealt with only at the terminals. There is plenty of room here for enlightened systems engineering. It appears out of the question to take the passengers in the 1980s to and from the airports by buses, let alone in private cars.

34. *Ground traffic safety*

There are experts who believe that nothing can guarantee complete safety on the highways short of remote control, independent of the driver (indeed, what else could guarantee against drunken drivers?). But electronic controls can fail too, and what will happen if the drivers, half asleep or asleep, are suddenly left to themselves? One can, of course, think of a fail-safe system, in which all vehicles stop dead when the electronic control drops out. The picture of miles of helpless cars is not very appealing.

There may be no safe remedy against drunk or doped drivers, but something can be done against the far more frequent danger of drivers falling asleep at the wheel. One could also think of continuous radar control of speed by automatic devices; and also of devices that automatically take a photograph of the car if it comes within safe braking distance of the one in front. It would, of course, be too expensive to stud the highways with such devices from end to end, but even a few in unknown locations might have a salutary effect.

Driving in fog presents special problems. There exists an ingenious system (RCA); a road in which green, amber, or red lamps sunk into the road surface indicate whether there is a car in front within a certain distance. It has been judged too expensive.

Small solid-state lasers which emit fog-penetrating infrared light are now available, and also small, solid-state detectors, which can indicate a car in front. This last solution, much less expensive, has a better chance of being adopted.

35. *Air traffic safety*

This, too, is a promising field for inventors. There are at present scores of near misses per day in the United States, fortunately with far less collisions. An ingenious new system (*Eros*, Westinghouse) will dictate to the pilot evasive action, ensuring, by a time division system, that other pilots will not be disturbed by the signal. After many premature attempts, it appears that instrument landing in poor visibility is now also safe. But as air traffic doubles in about seven years, the traffic problem is likely to remain acute. Even elementary problems remain to be solved, such as the tower-to-pilot communication. Usually, the intelligibility is so bad that it is a high credit to the refined senses of the personnel that there are so few misdirections and misunderstood directions. Better systems must be created before the traffic density increases to a dangerous extent.

2.5 Communications

Communication is, and I hope will be, increasingly in competition with transport. Business travel, salesman to customer, director to branch, is a considerable fraction of all travel. Business visits across the Atlantic in luxury liners were not sharply distinct from travel for pleasure, but the pleasure has gone out of business trips with the advent of the aeroplane. They are now strictly for communication *viva voce* and *viva faciae* in cases where the telephone is inadequate. But while the businessman may have a few hours' holiday away from his desk in a comfortable aeroplane seat, not even that much can be said for the commuter. His daily drive to and from his desk, watching traffic lights and the car in front of him, ought to be considered, by any reasonable standard, a waste of time and of nervous energy. Of course, man is far from being a rational animal, and hundreds of thousands of commuters are quite satisfied with spending one, two, or three hours at the wheel; because then at least they know what to do, and the danger of accidents perhaps even adds a little spice to it. I believe, though, that the large majority would welcome a world in which travel would be mostly for pleasure.

36. *The telephone*

There is nothing wrong with the modern telephone except that there are not always enough lines. More than one English firm which, following the advice of the government, moved its seat to a provincial place, had to go back to London because the lines were always engaged. If we are determined to stop the growth towards Megalopolis, ample lines must be provided, even if they are not commercially profitable. In the United States this can be done by tax relief; in the United Kingdom, where the Post Office is nationalized but self-supporting, by diverting public money to it.

37. *Wired television*

Wired television (also called 'Cable TV' or, misleadingly, 'Common Antenna TV') is highly desirable, if for nothing else, for taking the ugly aerials off our roofs. It is bound to spread because of (1) its high quality; (2) its freedom from interference; (3) its freedom from multiple-path reception; and (4) for its facility to be operated with 'pay-as-you-view' systems. (Television with aerial reception requires complicated scramblers for preventing non-payers from enjoying such programmes.) A choice of programmes could be ordered from the local centre through the telephone or through a simple signalling device built into the receiver, and could be paid with the telephone bill. In the United Kingdom, there are about a million homes wired up. In the United States, it is, for the time being, obstructed by commercial interests.

A wired television system will also be a great advantage in multiple-access computer systems which may lead ultimately to a nationwide data bank, because it would provide a wide-band channel in homes and offices.

38. *The Picturephone* *

The Picturephone of the Bell Telephone System has now been installed in Pittsburgh and other towns are soon to follow. Many years of research have failed to produce sufficiently economical band-compression devices; the commercial device has a screen of only $5 \times 5\frac{1}{2}$ inches, sufficient to show one face when operated with a wave-band of 1 MHz.

It is unlikely that compression ratios of more than perhaps 1 : 4 can ever be achieved, even with very complicated storage devices at the receiver end. If one would want to send a television programme through a telephone line, the receiver would have to be a computer hardly less complicated than the human brain. (It is only the brain that can achieve the miracle that with the *fovea*, which spans only half a degree, it produces a picture that gives us the impression of being sharp

* Registered Trademark, Bell Telephone Company.

over almost 180 degrees.) It is more likely that wide-band communication channels will become cheaper.

39. *Interoffice communication*

If it is only desired to see one's partner, a picturephone with 250 lines will do; but if it is desired to show full-page printed matter or technical drawings, the definition must be increased to at least 1000 lines, and preferably to 2000. Moreover, an equally good electronic camera must be able to view the drawing on the one side, the receiver screen at the other, which in turn, must be transmitted to the receiver screen of the sender so that both can point at the same place of the drawing and clear up doubtful points. The bottleneck here is that as yet there exists no electron camera with more than 1000 lines definition, and that wide-band communication channels are not available. In fact, helical wave guides or laser light pipes would be sufficient to carry the video communication of sizable towns, but so far, they have not proved commercially profitable. Here again, technical research must be backed by government action, to turn social benefit into commercial profit.

40. *Electrical transmission of newspapers*

An extension of the ticker tape, now almost a hundred years old; a modest realization is the telephone line with a telewriter. Modern teleprinters, as used in computers, can print faster than a man can read. If one wants newspapers or magazines with pictures and varied type, facsimile transmitters are available, of which the most advanced is, at present, the Xerox Telecopier II, which transmits documents through the telephone for office use. All this is much too expensive for the home, but it is in any case doubtful whether the home-reproduced newspaper would be desirable. It would require such a central organization that it would probably put out of business the local newspapers, and might represent a danger to democracy.

2.6 Computers and data processing

Data processing is practically synonymous with routine paper work, and also with office work, except (at any rate, for the time being) high-level decision-making and really creative work. As routine office work accounts nowadays in the advanced industrial countries for one-half or almost one-half of all working hours, the computer and the modern methods of data processing that it has made possible cannot fail to have a profound influence on almost every feature of social life. The computer has speeded up arithmetical operations by a factor of many millions. This has suggested to some that it may have an even greater influence on modern life than transport, which, with the jet plane, has added only a factor of 200 to the walking speed of a man. There are now about 70 000 electronic computers in the world, of which 50 000 are in the United States and 5000 in Britain, yet their influence is not easily perceptible to the naked eye. The office population has steadily grown in all countries in the last ten years. Parkinson's Law has proved, for the time being, stronger than the computer.

In this field, technological forecasting is easier than foreseeing the social effects. One must be forewarned by the fact that around 1950, some very able people had estimated that the whole computation work in the United States could be carried out by a dozen of the (still rather slow) electronic computers which existed at that time. The estimate for Britain was *two* computers. Never before have such able people underestimated a market by a factor of at least 10 000. One might easily fall into the opposite error by assuming that the tremendous technical development which can now be confidently expected will be absorbed by society with equal ease.

41. *Speed*

The time of a unit operation is the delay of the electrical signal from one circuit element to the other, plus the response time of a semiconductor device. This is approximately the reciprocal of the waveband which the device can handle. In the fastest computers to date, this unit time is about 10 nanoseconds (one-hundredth of a microsecond). Though computers are being progressively miniaturized, the delay is not likely to be reduced much below a nanosecond. One gigahertz can be considered also a reasonable limit for semiconductor devices; hence, the time per unit operation is now probably only one order of magnitude above the ultimate limit; nevertheless, experts expect a speeding-up of computers in the next ten to fifteen years by a factor of 1000. This can be achieved only by increased parallelization of operations. It must be considered though, that the programming of so many parallel operations might involve planning work comparable to the drawing up of very complicated PERT charts.

42. *Large-scale integration*

LSI is now being very actively pursued, for it means cheaper components, improved reliability, smaller size, and much lower power. Hundreds of components have already been assembled on silicon chips of the size of a coin, to be used both in the memory and in the logic of computers. Self-repairing computers are likely to be realized in which the failure of one LSI unit automatically brings a spare one into play. Subroutines are likely to appear in the form of hardware. It is claimed that the logic and the short-time memory of the largest present-time components can be ultimately fitted into a shoe-box. There seems to be little point in this so long as the peripheral equipment of computers is not correspondingly reduced.

43. *Memory*

It now appears that LSI memories with two-state semi-conductor devices will replace magnetic core memories in the new generation of computers. These will probably be in the form of plug-in memories.

In permanent stores for very high-speed computers, holographic memories promise the highest performance. A fine-grain photographic film (or a thin crystal) in holographic recording can store at least 50 binary (black or white) patterns in the area, which is sufficient for only one fully graded ordinary picture, with the same definition. Each of these can be called up separately by a laser beam of the right inclination. Ten million bits per cm^2 have been already realized. No photo-electronic device is fast enough to read these out in one microsecond. Large-scale parallel operation is therefore planned. The binary patterns are broken up into units of 8×8, each read out simultaneously by 64 semiconductor photo-diodes, and a very large number of these can operate, each on its own silicon chip, on one card, again in parallel. This way, it is expected to read out 10^7 bits in one microsecond. The varying deflection of a laser beam is effected by Bragg reflection on trains of high-frequency sound waves in liquids, glass, or crystals. Devices so far realized have 100 resolvable distinct positions or 'spots', but several hundred appear possible.

Work on erasable holographic stores also shows promise. 'Curie-point' devices based on the demagnetization by heating with a finely pointed laser beam of thin layers of materials such as bismuth–manganese and gadolinium–iron–garnet, and read out by means of the Faraday or the Kerr effect appear to be at present suitable only for binary stores, but even their capacity may approach one million bits per cm². Electro-optical crystals and 'Eidophor' devices (in which a cathode ray beam produces a profile on a liquid or thermoplastic substance) are also being considered as erasable holographic stores.

44. *Input*

Though the punched card still maintains its dominant position, there are persistent attempts to eliminate it by coupling the keyboards directly to the computer, or to replace it by direct reading of type-written matter, written on *any* typewriter; not only those special ones that are adapted to magnetic readout. There are two chief lines of approach to the problem of character recognition; the electronic reader and the holographic. In electronic fount-readers, the letter is projected on a matrix of photo-elements, shifted to a standard position, and identified by certain features, using a classifying electronic logic. These have progressed so much in the last decade that their price has come down by an order of magnitude to about $40 000. In the holographic system, the letter is simultaneously sub-jected to a great number of optical tests, by projecting the optical spectrum of the letter on a matrix of test-holograms. These will respond by emitting or not emitting a sharply focused diffracted beam on a matrix of photocells, signalling the presence or absence of a letter or certain features of it, from which an electronic logic decides the letter.

Hand-printed numerals have been read successfully by similar machines, and have been used for the sorting of letters and cheques. The reading of handwriting by machines still presents a very difficult problem, probably out of proportion to its importance.

Voice input to computers is a problem attractive by its difficulty. Numerals have been read successfully, but not for all speakers. It appears doubtful whether a machine simpler than the auditory cortex can ever achieve it; and even humans can misunderstand spoken numerals, while errors in computer input can be expensive.

45. *Output*

The typewriter terminal is still the most important, and where its speed fails, electrostatic printers, in particular cathode-ray printers, can take over. Voice output of computers is technically possible because not only numerals but also speech can be synthesized by tapping in succession a great number of small magnetic drums, each with its ready-printed phonemes or a combination of these. But this is not likely to be used except as a warning signal, or in cases where the operator's eyes are otherwise engaged, for the intake of the ear is much too slow. On the other hand, graphic output is of growing importance; in particular CRT (cathode ray tube) output, which will be mentioned again under 'Design'.

The paper consumption of modern computers is such that since only a very small fraction of the output is permanently stored an erasable output may deserve consideration.

46. *Time sharing*

This has advanced to the point where nationwide shared computer networks and data banks are being seriously considered. Logically, industry-wide networks would constitute an intermediate stage, but the difficulty of making independent firms cooperate is well known. It is more likely that great firms will establish their own time-sharing networks first. This may well go to the stage at which all records and documents of a firm are kept in the magnetic drums or discs or holographic records of a central computer, retrievable at a few seconds notice via a facsimile recorder or cathode ray tube.

The idea of extending such a central service to homes so that any information can be called up through the telephone is somewhat nearer to science fiction. The bottleneck is in the voiced input. Digital input (perhaps with a small typewriter, or at least a keyboard with numerals attached to the telephone) is closer to reality, and voiced output is by no means impossible. (But how will human nature take to messages starting with, 'This is a recording...?' Unless one can converse fluently with the computer, this is likely to drive some people mad.)

(IFF: regional data banks, 1973–1980–1985, experts 1980.)

47. *Engineeering design*

One of the most impressive of computer devices is the 'Sketchpad', a computer whose input and output is on the face of a cathode-ray tube. As an example, let the computer be programmed for linear circuits. The designer sketches in, with a 'light pen', a rough scheme of the circuit consisting of resistors, inductances, and capacitors, which the computer immediately corrects to a precise drawing with standard symbols. The impulse response and the frequency response of the circuit appear on the CRT face at the pressure of a button. The designer corrects the data until he is satisfied, than adds further circuits to it (if he wants to design a ladder network) and obtains the responses every time, however complicated the network. The Sketch-pad can be programmed also for plane structures, with the stresses immediately appearing once the loads are indicated, or for entire bridges.

Machines like this ought to eliminate the routine engineer who used to fill in design forms with data worked out on the slide rule or on a desk computer. They can even eliminate drawings, for the design data can be recorded on punched tape or magnetic tape which go into digitally controlled machine tools. They have already largely elimin-ated the hundreds of draughtsmen who in the 'lofting rooms' of aeroplane builders or shipbuilders used to draw out wing-covering sheets or ship plates in natural size on paper. They have also broken into the field of the highly trained designer. The computation of an aeroplane wing, with more than 1000 elements which would have cost hundreds of stressman-years (if the expenditure could have been afforded), now takes twenty minutes and the optimization a few hours.

The conclusion appears unavoidable that the computer will produce an 'aristocratic revolution' in the engineering profession. Only creative minds will be needed and competent programmers. Yet the engineering universities are turning out more and more graduates, the majority of which cannot become anything other than routine engineers.

48. *Software**

Most computer users merely supply data to be processed according to a programme already written; however, there is a large and growing need for new programmes. These are not usually obtained from first principles, but incorporate or utilize standard programmes already prepared.

Within the last ten years the provision of a range of standard programmes for a computer has become recognized as being as important as, and complementary to, the design of the computer itself. The programmes, now known as the 'software', are available to all users; they are called into play by the individual programmes that are written for specific jobs.

* For information on items 43–48, particularly the last one, I am indebted to Professor Stanley Gill, Director of the Centre for Computing and Automation, Imperial College, London.

Among the most important items of software are the compilers. These are programmes that direct the translation of other programmes from the language in which they were written to the form in which they can be obeyed directly by the computer. Many hundreds of languages have been invented but only a few are in widespread use; notably Fortran, Cobol, Algol 60, and the new and comprehensive PL/1. Nevertheless, there is still a need for ranges of compilers to meet various requirements of speed, efficiency, etc.

Software design and construction has become a new kind of engineering, and is practised by specialized firms. Some people believe that advances in programming languages and methods will so simplify the task of programming that the need for such specialists will dwindle. It seems more likely however that the spread of computers will demand more software rather than less, and that the search for more efficient ways of constructing it will call for even more highly trained specialists. Until about 1964, computer hardware technology advanced almost independently of the software; since that date the influence of software technology has become increasingly apparent, and it is now setting the pace of new developments in computing.

49. *Management information systems*

MIS is a vague but ambitious fashion word. It appears to be mostly used to impress managers that they are behind the times if they are using files instead of computer memories, or if they consult printed statistics instead of calling on data banks. The technical means are available, but if MIS means anything, it ought to mean that the new communication channels with almost unlimited capacity must be able to supply managers with better data for their decisions than those of which the previous generation could dispose. This, for the time being, is not a matter for machines but for human brains, for market researchers and economic forecasters, who, of course, can be assisted by machines; in particular by simulators. There is as yet no sign that economic uncertainty has decreased in the last decade, which saw the arrival of the computer, as compared with the previous decade.

However perfect the art of forecasting may become in the future, it must stop at decision-making. I fully agree with what Erich Jantsch said at the NSIA Symposium on Technological Forecasting, 18 October 1967:

All we know of human behaviour suggests that totally rational decision-making will eliminate the driving force of challenge and decision in the dynamics of human society; the rationalization which we are trying to achieve through planning is the *rationalization of the basis for action.*

50. *Information retrieval and associative computers*

Every computer associates in a primitive way when it connects an address with the contents of the store. This is sufficient for information retrieval whenever human minds have created a complete classification of the universe of discussion, as for instance in the universe of chemical substances. Such computers are the equivalent of a completely ignorant librarian, who can deal only with pre-classified material, but by their speed and by the vastness of information stored they can of course beat any human librarian. Where the classification is incomplete, the danger is that they overload the user. They can give him, for instance, data on a chemical substance which have been superseded by more recent work. A *selective* abstracting service, which is continually and critically brought up to date is therefore necessary even in fields in which good classification exists.

Selective information dissemination systems, in which every new paper is characterized by a few significant words, indicating the subject matter, and which are collated with the 'profile' of the user, are under active development. In the case of new material overloading is almost unavoidable. Even papers appearing in journals which have a good referee system can be judged for their importance or relevance only after several months, when they have been critically judged first by young experts, who have time to read journals, and then by the older experts, who have no time for it, and who rely on the younger people for information. But even an 'instantaneous critical system' of information dissemination, if such a thing were ever possible, would be insufficient in our time in rapidly developing fields, because by the time a paper appears, its authors are likely to have reached a further stage. Paradoxically, in our time of unparalleled rapid communication techniques, most scientists and technologists acquire their relevant information by verbal communication, by visiting conferences where new ideas are discussed in the lobby, or by visiting laboratories where they know that 'good' people are at work (a judgement which cannot be left to computers). In view of this difficult situation we can take comfort from the consideration that there is no need whatever to speed up the rate of scientific and technological innovation. It is already far too rapid for society to adapt itself to it.

A *self-organizing* computer, which could compete with the human brain in meaningful, critical abstraction of the input, and associate it with an equally critically pre-selected sample of relevant stored knowledge, is at present beyond our sights. Perhaps it will become possible only after we have learned to understand the human brain. But *pre-organized* associative computers are likely to appear gradually, well before the end of the century.* They are most urgently needed in medical statistics, for the rapid correlation of the side-effects of new drugs, and for the synergic effects of combinations of several drugs with other circumstances; instances so rare that no single doctor or even hospital could possibly collect significant data.

Another line in which progress can soon be expected is association with *incomplete* inputs. P. J. Van Heerden was the first to point out (1963) that holograms can restore a complete record from a fragment. This was generalized by H. C. Longuet-Higgins and by the author. It is not necessary to use an optical hologram; the mathematical process can be simulated on any computer. Such 'holographic' associative memories can not only restore a text from a fragment, they can also associate it with other relevant texts, but the association will have to be put into the computer by the designer. Multiple association, and associations with the consequences are also possible. Such computers may well give the impression that they are 'intelligent', while they are rather equivalent to a wise old man with a good memory.

51. *Automated credit, audit, and banking system*

This will be realized when nobody will pay with cash or even with credit cards which have to be verified by a clerk, but only with punched cards or magnetically or optically readable checks. These will be instantly verified by the bank of the buyer, credited to the seller, and reported to the central tax authority. Purchase tax can be automatically deducted; the rest of the tax will be computed on the basis of all

---

* We must not feel too superior to the machine, because it has to be told what to associate with what. The machine at least has no urge to make instinctive wrong associations. For at least fifty centuries men have associated human fate with the constellation of the planets at the hour of birth. To this day, according to Edward U. Condon, there are in the U.S. 10 000 people gainfully employed in astrology, while only 2000 are employed in astronomy. Nor is the computer inclined to associate old women with witchcraft.

financial transactions of the citizen. This is, of course, if not the naked society, at least the economically naked society. Advantages are that purchase and transaction taxes, which are likely to form in the future a growing fraction of public revenue, can be instantly and reliably collected; and also crimes for financial gain will be reduced to a minimum. The disadvantage is that it would make hundreds of thousands, if not millions, of clerks redundant.

52. 'Hierarchic' computers

The human mind can think 'hierarchically'; that is to say, on many levels of complexity, mastering the increasing complexity by finding laws and rules for each level. For example, chemistry can be conceived as atomic and molecular physics, from which only the concept and the laws of valency have been abstracted. (This is not the way chemistry has come into existence, but modern theoretical chemistry is doing this quite consciously, by a simplified wave mechanics.)

From atomic physics there is an ascending ladder of hierarchy through molecules, macro-molecules, cells, organs, organisms, to social bodies, each with its own laws, from wave mechanics up to the laws of psychology, mass psychology, economics, and the common law. Existing computers can be programmed in a higher hierarchy (for instance, in medical diagnosis and even in the criminal law) but without the ability of the human mind to eke out the incompleteness of the laws of one level by insight into the one below it. (For instance, physiology for medical diagnosis, psychology for the criminal law.) There is no fundamental reason why computers should not be able to achieve this in the future, but hardly without the help of very expert human programmers. For the present, it is hard to imagine that a computer could recognize, for instance, the symptoms of kidney trouble it has never seen before, or of a childhood trauma, by purely physical tests. Before computers reach this stage, *Gestaltpsychologie* must become a science instead of a collection of very interesting but rather baffling experiments.

This brings up the question whether computers in the near future will be able to beat human intelligence. John v. Neumann has given a very good answer to this when asked in a lecture if there were nothing computers could not do: 'Madam, if you can tell me *exactly* what it is that a computer cannot do I may be able to give you a computer that can do it.' The computer fails conspicuously when we cannot instruct it *exactly* in what it ought to see. The physician cannot say exactly what he sees in a 'kidney face', the art expert when asked for an attribution cannot tell exactly why he believes this picture to be a Picasso and not a fake. (Of course, neither the doctor nor the art expert are always right.) On the other hand, when we can specify the data and the rules exactly, the computer can do it, beating the human not only in speed but also in the complexity of problems it can tackle.

(IFF: computer to score above IQ 150, 1980–1990–2010, experts 1980.)

53. *Computers for the simulation of social systems*

This, again, is not a question of computer technology, but of human knowledge which can be fed into a computer. Improving this knowledge is a task of the very greatest importance.

The state of this art at present is probably best exemplified by Jay Forrester's model of urban growth and decay (*Urban Dynamics*, MIT Press, 1969). Forrester considers nine variables: three types of industry (from new to obsolescent); three types of housing; and three types of labour (from professional to underemployed) in a city embedded in an industrial country. In order to make the model work, he had to introduce laws connecting these with one another; for instance, the influence of the availability of highly skilled labour on the growth of new, progressive industries, of the availability of low-cost, subsidized housing on the influx of underemployed from other cities. Evidently, however plausible these assumptions, there appears much room for criticism; but his model receives strong *a posteriori* justification from the remarkable fact that the results are, to a high degree, independent of his assumptions, and also that they give a very realistic representation of historical examples of city growth over periods of 100 to 200 years. A most important general conclusion appears to emerge from attempts to manipulate this model. Almost everything that helps in the short run will be damaging in the long run. For instance, subsidies for remedying an acute shortage of low-class housing may produce such an influx of underemployed that new industries will be discouraged, taxes will increase, and after a time, the housing shortage will be worse than before.

This, of course, would not hold in a model of a whole country which keeps its door closed to immigration, but the lesson, 'short-term remedy—long term-deterioration', is probably a very general and important one for all democracies in which statesmen and administrators are elected for four or five years and are thereby forced to stress short-term remedies to ensure their re-election.

Where can the knowledge come from for improving models of social systems such as Forrester's? Only to a small extent from a study of history, which is essentially the record of *one* game played out in a myriad of possibilities. Moreover, modern technology presents us all the time with problems which are without parallel in history. A substantial improvement can come only the *very* hard way; from the gradual discovery of the laws of social dynamics. It must be admitted that one can easily be discouraged by the difficulties of the task. Statistical mechanics is a great achievement of physical science. It predicts successfully the behaviour of complex entities such as gases or liquids, based on the simple laws by which their atoms or molecules interact. But statistical mechanics has come up against almost insuperable difficulties recently when it has met the phenomenon of *collective interaction* in plasmas, where every ion interacts, not only with its next neighbours, but as it were, 'knows' what all the others are doing, and by this, produces instabilities which have, so far, defeated all attempts at controlled fusion.

This, however, is as nothing compared with the complexity of human societies where the individuals (or organizations such as firms or nations) try to *anticipate* what the others will be doing. (Think of the stock exchange as an obvious example.) However, the problem of stabilizing human society by long-range foresight is of such importance that nothing must discourage us. Unless we master the art of maintaining a stable world while preserving the maximum amount of human freedom, the alternatives will be either catastrophies—or the police state, which ensures stability by clamping down on everything.

2.7 Robots

54. *Man-controlled mechanical hands*

In their simplest form, these are manipulating devices which copy the motions of an operator. In a more advanced form, they are assisted by servo motors similar to the power steering of motor cars. The first form can perform manipulations in health-endangering conditions (radioactivity, heat, underwater); the second can also perform tasks exceeding human strength. The Danish inventor, Leif Sorensen, has constructed electronic-hydraulic remote control systems for earth excavating machinery and the like which save the operator from the discomfort caused by the tilting and the vibration of these machines.

A closely allied type of device is that of modern servo-assisted prostheses which enable amputated or limbless persons (such as Thalidomide victims) to work their artificial limbs with the muscles in the stump. Operating these with the action currents of nerves in the stumps is a desirable aim of these developments, but to the knowledge of the author, it has not yet been realized.

55. *Programmed mechanical hands*

The most advanced of these robots is the UNIMATE (Consolidated Controls Corporation, Bethel, Connecticut, U.S.A.) The inventor is George Devol. It has motions humanoid in form, simulating the waist, shoulder, elbow, wrist and fingers, with a great variety of 'hands'. The 'hand' is taken once, slowly, through the operations which it must perform; afterward, the speed can be set, and the robot will go through a series of up to 200 separate steps at any speed, and for any time. The steps are stored in a magnetic memory, the actuation is hydraulic. For instance, the UNIMATE can pick up an ingot from a rack or chute, put it in a die casting machine, pull out the still red-hot casting and place it in another rack, in an empty place, taking account of the castings which it has previously put in, until the rack is full. In this and many other applications, it achieves a consistency beyond that of the best human operator, 24 hours per day. Workers and even labour unions in the United States have, surprisingly, taken kindly to this robot. They consider it as a willing slave, not as a blackleg.

56. *Robots with sensory feedback*

Some of the UNIMATES employed in diecasting have a first vestige of a sensory feedback by an infrared system which checks the temperature of the casting and if it is not right, the robot goes into an alternative sequence of operations. Progress beyond this point is likely to be extremely difficult. 'When one tries to design a robot, one gets proper respect for the intelligence of a moron.' (J. F. Engelberger, co-inventor of UNIMATE.)

Robots with sensory equipment have been designed in MIT and in Stanford University, in the Stanford Research Center, but though they are coupled with digital computers, their performance to date is still very elementary. The 'electronic housemaid', a project of Professor M. W. Thring (Queen Mary College, London), though it can perform motions such as climbing steps and clearing a table, is still weak in its sensory equipment. One can say that it is incomparably easier to design a computer for solving a wave equation beyond the reach of the best analyst than to design one which will pick up and empty ashtrays, because ashtrays come in so many shapes. Here we meet again the weakness of computers in the recognition of 'universals'.*

(IFF: electronic housemaid, 1990–2000, experts 1990.)

57. *The mechanical typist*

A device that types out the spoken words, phonetically if not orthographically, is a favourite of SF, and also, regrettably, of some electronic trade journals. It is extremely difficult, which is just as well, because if successful, it would make millions of typists redundant.

58. *The mechanical linguist*

A machine which translates from one language into another. Word-by-word translation has been achieved long ago; also translations in which alternatives are indicated in case of doubt. Meaning-preserving translation has been achieved, so far, only by annotations of an operator who knows the original language. No machine is yet in view which could preserve the meaning and put it in the second language in a grammatically correct form. Owing to the complicated and often illogical structure of 'natural' languages, such a machine would be hardly less complicated than a human brain. This is fortunate, because if any such device were successful, it would discourage young people from becoming linguists. Instead of spending money and effort on this folly, it would be preferable to have bilingual schools for people at the age of six to twelve when they can learn two or three languages almost without an effort. They could then travel in foreign countries as if they were their own and enrich their minds with the literature and poetry; thus, the 'mechanical linguist' is one of those, fortunately rare, cases in which technological 'progress' is head-on opposed to what deserves the name of civilization or culture.

(IFF: 'laboratory operation of automated language translators

* For information on items 54–56, I am indebted to Mr. J. F. Engelberger, President, Consolidated Controls Corporation, Bethel, Connecticut, U.S.A.; Mr. D. F. Cappell, Manager, GNK Machinery Ltd., Wolverhampton, England; and Professor M. W. Thring, London.

capable of coping with idiomatic syntactic complexities', 1980–2010, experts 1980.)

2.8 Automation

I want to discuss under this heading *all* technological innovations which result in increased production at decreasing human effort. This includes mechanization and rationalization, as well as automation in the restricted sense, which means production and process control with feedback. It is true that control with feedback has added a new dimension to automatic machinery by making it capable of operations which formerly could be performed only by skilled or semi-skilled workers. However, it makes little difference for the manufacturer whether he increases his productivity by one type of machine or another, so long as it is effective; nor for the worker whether he is made redundant by a device with sensory feedback or without. Even those authors who, like John Diebold, strongly insist that an entirely new era has been opened up by feedback or 'cybernetic' machinery, cannot refrain from talking of 'Detroit Automation', though the transfer machines (in fact used by Morris Motors, in 1923, some twenty years before Ford), have no feedback. It may well be that in the long run machinery with feedback and backed by computers will have the greatest social impact, but until now simple mechanization (such as containers) and rationalization (such as the streamlining of office work, carried out by Marks & Spencer) has had the greatest effect on employment.

Robots of course also belong to automation, but I have singled them out because they are more closely related to the computer family.

The social impact of automation is a highly controversial subject.* Nobody can deny that it is capable of liberating mankind of all, or almost all monotonous drudgery, of mining with the pickaxe, road

---

* For a withering all-out attack on automation and all its works, supported by an enormous amount of documentation, see Ben B. Seligman, *Most notorious victory—Man in an age of automation*, The Free Press, New York, 1966.

For an optimistic, soothing appraisal of automation, also supported by much statistical material, see Charles E. Silberman, *The myths of automation*, Harper & Row, New York, 1966.

The fact that two serious American students of automation could come to such contrary conclusions contributed to my decision to use the excellent German statistics of Dr. Friedrichs rather than American data.

building with spade and wheelbarrow or of the stultifying, mind-numbing work at the conveyor belt. On the whole the reports on the mental health of workers in automated factories are very satisfactory. But what happens to those who have been made redundant? Technologists and economists usually have two sorts of answer to this. One is that *so far* everything has gone well, the redundant production workers have been absorbed by the service industries and by the offices. This is perfectly true, but it reminds one of the optimist who falls from the tenth storey, and by the time he reaches the third notes with satisfaction that 'so far it has gone well'. The other answer is hardly more satisfactory. It is that the redundant workers can be used in new production jobs, to produce more wealth. This is fine so long as other, expanding industries are waiting for them, but are they?

A good idea of what technological change has achieved so far can be obtained from the excellent statistics compiled by Dr. Günther Friedrichs of the Industriegewerkschaft Metall (the trade union of the German metal industry).

Production data for selected industries in West Germany, showing the changes between 1958 *and* 1966*

Industry	Production volume (%)	Working time per employed %	Productivity per working hour (%)	Number of employed (%) absolute
Mineral oil	+ 111·0	− 3·8	+ 220·7	− 31·6 − 4100
Tobacco	+ 63·8	− 9·6	+ 223·7	− 44·0 − 29 000
Textiles	+ 48·2	− 7·6	+ 86·2	− 13·9 − 87 000
Fine ceramics	+ 27·6	− 9·2	+ 59·1	− 11.7 − 11 000
Coal mining	− 11·3	− 8·3	+ 59·8	− 39·5 − 254 000

The last row is particularly interesting, because it shows that productivity can increase even at decreasing production. In the period 1961–6 there were in West Germany 26 industries in which the number of employed had dropped (by a total of 370 000) while the productivity

* Dr. Günther Friedrichs, 'Technischer Wandel und seine Auswirkungen auf Beschäftigung und Lohn' in *Lohnpolitik und Einkommensverteilung*, Berlin, Duncker & Humblot, 1969.

had increased without exception, 4 industries in which the number of wage earners dropped while the total number of employed increased owing to more employees, and 12 in which both types of workers increased (though wage earners always much less than employees), with an increase of 423 000, giving a total increase of a modest 70 000 or 0·8 per cent in the number of all employed, at an increase of production volume of 25·4 per cent in five years. It appears from this that the time cannot be very far off when the number of production workers in Germany will start dropping, as it has done in the U.S. for more than a decade. It may be noted that during this period automation in Germany was going rather slow, owing to fear of unemployment, lack of capital, and a cautious financial policy, which held back domestic consumption.

The accelerating rate of total productivity in Germany, over a period of more than a hundred years, is shown in an interesting table.

Total German productivity 1850–1966

Period	1850–1913	1925–38	1950–9	1960–6
Annual increase per worker (%)	1·44	2·26	5·41	4·06
per working hour (%)	1·73	2·88	6·7	5·0

The rate of increase had slowed down a little in the last period, but only in comparison with the previous period, 1950–9, in which the almost destroyed German industry had to be built up again. It may be premature perhaps to see in this an early sign of the development which has taken place in the United States, where the productivity has actually *dropped* in the last months of 1969, after an almost uninterrupted increase of about 3·5 per cent per annum since the end of the Second World War.

If one looks only at the past and the present, complacency might appear justified. The total employment is at a high level; in Germany and in Britain 47 per cent, in the U.S. 35 per cent of the population are gainfully employed. But the distribution has changed in a significant way. In Britain and in Germany the manual workers are now a little less than half of the labour force, in the U.S. only one-third. The rest are service and office workers.

Will the offices be able to absorb the redundant production workers in the future, as they did in the past, thanks mainly to Parkinson's Law? Certainly not if rationalization and computerization will be applied to them as relentlessly as they were applied in the factories in the last two decades. On the other hand, after the terrible experiences of the thirties, and after the *belle époque* of the twenty five years after the last war, no country can tolerate large-scale unemployment. Popular anger would sweep away any democratic government which would allow it to rise above 5–6 per cent, perhaps even earlier. In this dilemma technologists and economists of good will do not see any other way than to let technological change go on (in moderation), and let the production be absorbed by increased popular consumption. There is no such thing as 'saturation'! This view may be justified for perhaps ten more years in the Western countries which have not yet reached the American standard of living, but in the U.S., the most advanced country, the crisis cannot be far off. At the time of writing the U.S. are maintaining their unemployment on a fairly low level by a large-scale war in Vietnam, by keeping $2\frac{1}{2}$ million young men under arms, and by using more than 10 per cent of their GNP for armaments. Besides, at the modest rate of productivity increase which they have maintained since the war it would take a 4·5 per cent rise per annum in consumption to keep the unemployment at the present level. (Estimate of the 1966 National Commission on Technology, Automation and Economic Progress.) By A.D. 2000 this would mean a more than fourfold GNP, with perhaps $20-30 000 spending money per family, and this is hard to imagine. Something will give. In the best case it will be the rate of technological progress, in the worst case it may well be the whole 'achievement-oriented society', as the revolutionary fraction of American youth would like to believe.

I have written this long preamble to the innovations in automation, because, apart from war inventions and perhaps bio-engineering, this is the field in which a foresight of the social consequences is most badly needed, and in which the technologist must tread most cautiously. Innovations are justified if they eliminate monotonous, degrading work, below the intelligence level of the worker, provided that more satisfying jobs can be found for him. They are also justified under the pressure of the necessity to maintain the balance of payments in a competitive world, coupled with the pressure of the 'rising expectations'. Needless to say that better social mechanisms will be needed than those which we possess at present if reasonable progress is to be maintained. The trade unions very often press for higher wages, while opposing the innovations which can provide these without damaging the rest of society. They are also anxious to retain their membership, and are not much in favour of retraining programmes, which are indispensable in a progressive economy. Nevertheless, one must admit that much progress has been made in the understanding between employers and workers since the days of the Luddites, and that automation has generated rather more hope than fear in the wage-earning classes. How else could they expect to get more for less work?

59. *Flexibility*

One can conceive a transfer street which is suitable for making cylinder blocks and Wankel engines, but hardly one which can also manufacture electric motors and cash registers. On the other hand many industries have operations in common. For instance the casing of an electric motor requires only ordinary machining operations. One can therefore imagine automated factories, with transfer machinery flexible to the limit of economy, which carry out common processes for all industries in a district or even in a country.

In the manufacture of small parts the present peak of development in automated transfer machinery is probably the '24' system devised by Theo Williamson for the Molins Machine Co., London. Molins are makers of cigarette manufacturing machines, and require thousands of different small parts, in thousands, not in hundred thousands. Batches of these are fixed on pallets by girls in an 8-hour shift, and these move automatically, through 24 hours a day, through a street of computer-programmed machine tools.

60. *Semi-automation of repairs*

It is a common complaint in our civilization that the customer can buy, at reasonable prices, highly sophisticated, automation-produced 'durables' such as television sets, refrigerators, washing machines, motor cars, but when something goes wrong with these, he is at the mercy of often rather ignorant and uncaring servicemen or garage mechanics. Many middle-class men, especially those with technical education, have therefore become skilled repairmen, but one cannot expect this of everybody.

This is one of those cases when there is a real spin-off from military developments. During the 1939–45 war highly complicated electronic apparatus had to be maintained in perfect working order by not very highly skilled personnel. These sets were therefore fitted with a great number of test points to which a cathode ray oscillograph could be connected, and the personnel were given manuals so that they could instantly diagnose the trouble from the CRO patterns. This practice has not spread to television sets and the like, in which even a small 'unnecessary' extra cost is considered as unbearable. Consequently customers often put up with almost unbelievably poor TV pictures rather than call in the serviceman, whom they mistrust. They would trust him if the serviceman came with a test set, and would demonstrate to the customer from the CRO pattern and the manual that this or that small component has to be replaced (instead of saying that 'the tube has to be replaced' or simply that 'the set has to go to the factory').

The same practice could be established without difficulty in testing the electrical equipment of motor cars, with some thought probably also to mechanical defects, perhaps by displaying vibration and sound patterns. If customers would more often take their cars for checking, there would be less accidents. The insurance companies have an interest in promoting improved and inexpensive checks.

61. *Automation in the home*

Automatic washing machines are now developed to high perfection. Ironing has become less important with the spreading of drip-dry fabrics. Laundries possess very efficient ironing machines, which do not imitate the motions of the human hand, but exert approximately hydrostatic pressure (which does not crush buttons) on the hot, stretched fabric. But as several such machines are required, one for each part of the garment, the introduction of such machines in the home is neither likely, nor particularly desirable.

The modern kitchen, with its programmed stove and dishwashing machine, is now so highly automated that the next step, putting away the dishes, would require a very costly change. The 'electronic housemaid', who could lay the table, clear it, dust the furniture without breaking china, and sweep or vacuum-clean floors, around the furniture and in all corners, is even more remote. It is more likely that when the population become richer, they will take to eating outside, and leave the cleaning of the home to an army of specialists, well paid and socially respected.

62. 'One-off' machines

Automation is a natural solution for mass production, but with the arrival of the 'cybernetic' machines, not only SF writers but also technologists have thought of machines so highly adaptable that they could produce 'individual' goods. One of these machines, for 'bespoke tailoring', may not be far off. It is perfectly feasible to instruct a mechanical cutter with individual measurements. Taking these measurements with another machine is also possible, though hardly necessary. (Dean Swift has already gone a good step further in his Laputa, where Gulliver's measurements were taken with a theodolite, and the result was a singularly badly fitting garment, because there was an error in the computation.)

I would not object to mechanized bespoke tailoring, because a well-fitting individual suit is not exactly a work of art, but I would strongly object to one-off machines which would produce artistic objects. One could quite well conceive machines which produce, for instance, cut glass, not only by executing a design, but by introducing variations with a foreseeable pleasant or interesting effect. Here the machine would infringe the human sphere. The artistic crafts must be revived, because men and women want to take pride in the products of their skilful hands, as shown by the spontaneous spreading of the 'do it yourself' movement. This revival is not only economically possible, but psychologically necessary in the post-scarcity epoch. The machine could do it very well, but it must not be allowed to transgress into fields such as embroidery, Brussels lace, hand-painted china, hand-cut glass, or individual bookbinding.

63. Time saving in the home by waste

The 'throwaway revolution'. This is not automation in the home, but a fairly straightforward consequence of automated industry producing cheap disposable goods in billions. Paper plates, cups, bottles, containers have long been disposable; these are now joined, according to a recent report, by 'everything from bikinis to men's blazers, nightwear to student's gowns, curtains to bathmats'. In 1969 about half a billion worth of disposable goods have been sold in the U.S. and this is expected to quadruple by 1980. Long before that time the disposal of these 'disposables' will have created a problem. Will the boredom in a society in which there is consumption but no possession be more or less than it is today? This is a question that belongs to the social problems created by mass production.

2.9 Education and entertainment

Education and entertainment are best discussed together, and not only because the technical means are the same in both cases. This is not new; books have always served for education and for entertainment as readily as their latter-day successors, films and television. But the two will have to come even more closely together in the era of mass-education and mass-culture. In the past, when perhaps 10 per cent of the population were educated to high-school standard and 4 per cent or less to university standard, education was a valuable social distinction. It was worth while for the select to work hard, and to absorb knowledge in ways which gave pleasure only to a minority of a minority: to the exceptionally gifted. But once education is spread over the vast majority, it cannot *ipso facto* confer distinction on all. Why then should those work so hard, that have less good brains, which do not naturally appreciate the intellectual and artistic treasures of our civilization? They are well aware of this. In the United States, where now about one-half of an age group enters the univeristy, surveys have shown that almost one-half of the students give as their reason for 'studying' the opportunity to 'change society'. This is probably an honest statement only for a small fraction; for the rest one suspects that they rather want to enjoy campus life for a few years before bending their neck to the yoke of a job. However this may be, we have reached the stage when the medicine of education must be given with a spoonful of sugar. Better then marihuana or LSD—and no learning.

64. *The television of the future*

All-wall, three-dimensional television, in which the viewer is in the midst of the happenings, is a favourite of science fiction. It could be undoubtedly a very powerful means of education. (It could be used also for maintaining a fantastic level of inanity and imbecility, as brilliantly described in Ray Bradbury's *Fahrenheit 451*.)

Coming down to the level of reality, cathode-ray tube screens larger than about 80 cm (32 inches) in diameter ought to have armour-plate thickness; a screen of a modest 125×100 cm (50×40 inches) would weigh about 200 kilograms. Television in such sizes will become possible only with solid-state devices. This is a long-standing dream of electronic engineers ever since the first practical electro-luminescent materials appeared, around 1950. But these have developed only little in the last twenty years, and there are good theoretical reasons which make a breakthrough very improbable. As they are now, they can yield the brightnesses to which one is used in television only with continuous excitation, unlike the phosphors in cathode ray tubes which receive their supply of energy in about four-millionths of the total time. Any electronic engineer can construct a transistor circuit in which a scanning signal with a duration of only a fraction of a microsecond opens a gate for a proportional flux of energy to last from one scan to another, but to make these at a reasonable price is a different matter. Even present-day standards require a quarter of a million independent luminous points (colour television three times more), and at *one cent* apiece this is $2500 or £1000, with a (still modest) screen of 125 × 100 cm about eight times more.

The wide waveband required for such large screens is a relatively small obstacle. It could be supplied in the UHF waveband, even better by helical waveguides or by laser channels.

Three-dimensional television takes us further into the realm of the impossible. *Stereoscopic* TV could be realized at present screen sizes with a doubling of the waveband, with rather sophisticated cathode ray tubes, a grade more complicated than colour tubes, but it would not be worth while. In the experience of the author, it gives the impression of a puppet theatre; by adding one touch of reality one becomes more conscious of the lack of another: real size. *Real* 3D, where the object, seen from different positions, shows different aspects, adds a further large factor to the waveband, and in the sizes in which it would become realistic and interesting one would have to contemplate wavebands of 1000 MHz or more.

Waveband-saving compression of visual information is in principle possible, but in order to make it effective one would have to add a computer on the receiving side hardly simpler that the visual cortex plus the *Gestalt* memory in the human brain.

The conclusion is that with 625 lines and 6MHz bandwidth popular TV has reached its maturity, for a long time. There is of course no obstacle to producing 1000-line transmissions for special purposes, such as office communications.

(IFF: 3D television, used routinely for entertainment, 1980–1990, experts 1992.)

65. *Cassette television*

It has been already mentioned, under section 37, that wired television offers a way for the viewer to choose his programme from a limited selection, under a 'pay as you view' system. 'Cassette television' is the generic name for the new systems, which enable the viewer to select his visual entertainment or instruction from an almost unlimited stock, comparable to the stock of audio (disc or tape) records. There exist, or are about to be developed, three such systems.

(1) EVR or Electronic Video Recording, developed by CBS Laboratories, Stamford, Conn., U.S.A. The record is a microfilm, with 50 (in the U.S.A. 60) frames 1·25 mm high and 3·4 mm wide per second for monochrome, two such frames side by side for colour, one for the luminance, the other for the chrominance signal. The master film is recorded *in vacuo* with an extremely fine electron beam, modulated in such a way as to pre-emphasize high frequencies and to compensate the imperfections of all stages up to the final display, which gives a perfect grey scale. The release films are contact-printed from the master, at high speed. In the player the film moves steadily, not intermittently, and is scanned with a flying spot from a CRT. The light is collected by a photomultiplier (two in case of colour) and the signal is fed into a television set. Unlike in a projector, the film can be arrested at any frame, without danger of burning it, which is especially important in instruction.

(2) The SONY videotape recorder, in its popular form. This is a magnetic tape recorder of the 'helical scan' type, in which a wide tape is wound in the form of a long-pitched helix around a rotating capstan which contains a magnetic head. One full rotation gives one frame. Suitable for monochrome and for colour. The release tapes are produced by magnetic contact printing from a master. An electron camera is available for amateurs to make their own programmes.

(3) The RCA holographic SELECTAVISION system. This system is not yet fully developed. The record is a film, not photographic but embossed with a fine relief; a 'phase' hologram. This is essentially a Fourier transform off the frame which is to be reproduced. The master hologram is produced by exposing a photographic film simultaneously to the frame, illuminated with laser light, and to a reference beam, from the same laser. No lens need be used in this process if the reference beam has a focus in the plane of the frame. ('Lensless Fourier hologram'.) The master film is then bleached, so that the density is converted into a fine relief, and a galvanoplastic replica is made of it, which is then used to emboss the thermoplastic release film. In the

player the film is illuminated with laser light, and the reconstructed frame is caught on a vidicon tube, which feeds the display tube. In this system no synchronization is needed. When the film is steadily moved, every point of the reconstructed scene changes continuously from one frame to the next. Colour is also in preparation.

It is not to be expected that TV displays will entirely oust the film in instruction, though they have the advantages that one player can serve any number of TV display screens, and that (in individual use) the record can be arrested at any frame.

66. 3D cinema theatres

The film as a video record comes into its own in stereoscopic and 3D displays. Stereoscopic movies had short-lived vogues in the past, first as anaglyphs (glasses with two complementary colours), later as projections with polarized light, which had to be viewed with polaroid spectacles. A stereoscopic theatre in which no glasses had to be used was first realized by the Soviet inventor, Semyon Ivanov, immediately after the war. The screen was a modified Lippmann lenticular screen, with conical lenticules, which converged at a point below the screen. A double projector was used, which threw on the screen one picture for the left eyes, one for the right eyes. The Lippmann screen sorted these out, by making the left eye pictures visible only in a number of wedge-shaped strips, which converged to the same point as the lenticules, and similarly for the right eyes. The system was imperfect, because the width of these viewing zones could be made equal to the normal eye spacing only at a certain distance from the screen. The theatre could hold 180 viewers.

Ivanov's system, a very respectable achievement, represents the limit of what can be achieved with the orthodox means of optics; lenticules, prisms, mirrors. Further progress is possible only with a holographic screen. A hologram automatically solves one half of the problem to be achieved by a stereoscopic screen, which is making visible the picture projected by one projector only in a system of viewing zones (strips) in the plane or planes of the audience. One only has to show to the photographic plate simultaneously the projector (a point source) and the viewing zones, both in coherent light. But the second half of the problem, the condition that the left picture shall be invisible to the right eyes and *vice versa*, can be solved only with the 'deep', 'volume', or reflecting type of holograms, initiated by the Soviet physicist Yu. N. Denisyuk in 1962. These have the directional selectivity required for separating one picture from the other; moreover they can be illuminated with ordinary (not laser) light, and show the images in natural colours. Such a holographic system is now under active development.

67. *Wide-angle stereoscopic projection*
Leonardo da Vinci, and many imaginative writers after him, dreamt of a picture which surrounds the viewer from all sides, so that he is in the midst of the happenings. It has been mentioned before that such a system, with the viewer freely moving, comes up against a formidable information-bottleneck. But if the viewer keeps still, or if he moves only a little, one has to present him with only twice the information which he is capable of taking in with one eye, and this can be accommodated in two 35-mm films. A device which can accomplish this consists of a projector with two wide-angle ('fisheye') lenses, the same as the objectives through which the films were taken, so that the distortion is compensated; every ray goes back in the projection where it came from in the taking. The screen is almost hemispherical, and sends back every ray from the left-eye projector into a strip sufficiently wide to cover the left eye of the viewer who sits below the projector, and similarly for the right eye. This system is under active development. It ought to be a teaching device of unmatched power.

2.10 Space
Three years before it was achieved, Lord Bowden called the plan to put a man on the Moon 'the largest, most extravagant and most highly organized system of outdoor relief'. There is of course much truth in this. On the other hand one must admit that, apart from giving work to the overgrown heavy and electronics industries of the U.S. and the U.S.S.R., and a comparatively harmless outlet for the ingenuity of an army of technologists who could pit their brains against the merciless hostility of Space instead of against the equally merciless hostility of the arms industry of the opposite camp, the Moon landings have given a wave of happy excitement and sense of collective achievements to the American people, with a remarkable resonance in other countries. What is probably more important, they have given once again a proof of what can be achieved once a dream becomes a *project*, and can be handed over to an army of talented and imaginative technologists with a single purpose. It ought to be a

model for social projects on a grand scale, once unity of purpose can be achieved.

After the success of the Apollo 11 and 12 projects the enthusiasm in the U.S. has visibly declined; NASA appropriations have been painfully cut and more than 50 000 workers have been discharged. A revival is likely only after a Soviet success. The scientific results of the Moon expedition were not exciting enough to arouse popular demand for a landing on Mars.

68. *Space platforms*

Their feasibility is not in doubt, but they could be justified only by increased Earth–Moon traffic. There is a strong suspicion that Soviet technologists are pushing this line because of its alleged military importance, and because support from the military is even more important for them than for their American opposite numbers.

69. *Photon rockets*

A very controversial subject. The nearest star, Alpha Centauri, is 4·3 light years away, and it is claimed that while chemical rockets could not get there in hundreds of thousands of years, photon rockets could make the return trip in just over twelve years. Against this stand the very careful calculations of the Hungarian physicist George Marx, who has proved that the efficiency of rockets *of any sort* decreases so rapidly at only 10 per cent or so of the velocity of light, that a manned rocket could make this journey in forty years only if it had an initial mass of the order of that of the Earth, even if it converted mass completely into light by Einstein's equation $E = Mc^2$. In other words, exploration of space beyond the planetary system is a dead-end occupation.

70. *An observatory on the moon*

The most important scientific result of space research to date is the discovery of X-ray stars, and others are likely to follow. But by far the most exciting discovery would be the detection of intelligent beings in the universe. There is a chance, nobody can say exactly how strong, that a radio observatory at the other side of the Moon, screened from the radio noise of the Earth, may detect cosmic radio signals, and prove that 'we are not alone'. *If* such a source were established, it might be possible by an effort comparable to that which has gone into the Apollo projects, to beam sufficient energy towards that source to establish communications with those beings. It must be admitted that a reply is not likely to come back in much less than a hundred years.*

Against these somewhat sanguine expectations can be set the sober estimate of E. U. Condon. Starting with a reasonable estimate of inhabitable planets (probably only a few within fifty light years), assuming that it is a matter of chance when life has started within a billion years or so, and making the (alas) reasonable estimate that a civilization is unlikely to survive a hundred thousand years after the discovery and exploitation of nuclear energy, he comes to the conclusion that it is extremely unlikely that *two* high civilizations could exist *simultaneously* within a radius of fifty light years.

(IFF: discovery of information proving the existence of intelligent beings beyond the earth, 1985–2025–later, experts 'much later'.)

2.11 Ocean research and exploitation

Only mining will be discussed here; the exploitation of the oceans for food will be left to the next chapter. According to some reasonable estimates the amount of nickel and copper mined from the ocean floor may be comparable to that mined on land within 10 to 15 years.

71. *Underwater robots*

At the present primitive state of robots with sensory feedback that could carry out work 'on their own intelligence' it is reasonable to think only of robots that are under the permanent control of a man on the surface, who sees exactly what the robot 'sees'. The 'robot' need not be of course anything man-like; it can be a collector of manganese nodules, a mining machine, or an underwater cultivator. More man-like robots could also carry out underwater repairs on gas rigs.

(IFF: 1980–1990–2000, experts 1985.)

72. *Underwater vision by sound*

In murky water ordinary vision may be limited to only a few inches. It is an old idea to replace light by ultrasound, with waves short enough for accurate vision. (Sound with a frequency of one megahertz has a wavelength of 1·5 mm, penetrates to a depth of the order of 10 metres, and with a viewing system of 10 cm diameter would give a resolution of 1 cm at one metre distance, 10 cm at 10 metres.) Development was held up for a long time by the lack of the equivalent for sound of a photographic plate or of an electron camera. (The

---

* Walter Sullivan, *We Are Not Alone*, McGraw-Hill, New York, 2nd edn, 1966.

Sokolov tube, an electron camera with a piezoelectric screen which translates sound into electric charges and thereby into a TV picture, invented in 1935, has been developed in the meantime by several authors to the limits of its capability, but gives at best very indifferent pictures.) It is also almost impossible to construct sound lenses of the required perfection. This second obstacle is eliminated in the schemes of acoustical holography, on which much work has been done since 1967. In holography no lenses are required, it is sufficient to expose to the sound an equivalent of a photographic plate, simultaneously with a 'reference wave' of the same frequency. The early attempts have led only to rather low-grade sound pictures, for two connected reasons. First, the equivalent of a photographic plate could be realized only by matrices of hydrophones. Second, though a 100×100 matrix of hydrophones, with their amplifiers, constitutes an impressive piece of electronics, it is still far from giving a 100×100 image. This is due to the phenomenon known in light holography as 'laser speckle': the excessive noise in images formed in coherent radiation, which reduces the effective resolution of an optical system by at least a factor of ten.

More recent attempts, still insufficiently tested at the time of writing, promise to overcome this difficulty by replacing the hydrophone matrix by a thin reflecting membrane, and enhancing its insufficient sensitivity by phase modulation of the illuminating light in synchronism with the sound wave.

The foregoing remarks apply to *simultaneous* vision by sound. In cases in which there is enough time left to put together a composite picture the difficulties are considerably less. Excellent sound pictures of the ocean floor have been obtained by 'side-looking Sonar' and similar scanning methods.

2.12 Peace inventions

We have now enough war inventions to exterminate the whole of mankind several times over. This will not of course stop their development, in the name of 'defence'. At the time of writing there is great concern in the United States about the Soviet SS 9 missiles, with their 25 Megaton, triple delivery warhead, and there was no time in the last twenty years in which the great powers have felt safe on top of their pile of armaments. This insane race is likely to continue, and also to extend to smaller countries, but I will not darken these pages by adding 'cheap Doomsday machines' and the like to the list of innovations.*

Are there no peace inventions? In fact, ever since the atomic scientists have acquired a guilty conscience, some of the best minds have been desperately searching for them. There was first the suggestion of contaminated nuclear fuel, with a contamination that would stop the fast nuclear reaction, and prevent the making of atomic bombs. Especially ingenious was the suggestion by Louis Sohn and Hans Bethe to overcome Russian touchiness about inspection. Each country is divided into, say, twenty regions. Lots are drawn, and one region is inspected, which still leaves 95 per cent of the 'surprise potential' untouched. But of course all these, and many similar, proposals were rejected in the name of 'national sovereignty' and similar cheap and dishonest excuses. There is only one peace invention which really works—so far: the artificial satellite.

73. *Inspection from satellites*

At the time of writing an unknown number of satellites from both sides are circling the globe about once in ninety minutes, taking aerial photographs and transmitting them to the ground by television methods. They are officially labelled 'weather' or 'geographical survey' satellites, and some really are, though of course all of them are capable of detecting missile sites and atomic factories if they are of a sufficient size, and cannot be sufficiently camouflaged. Nevertheless, by an unwritten and unspoken agreement, not one of them has yet been shot down, though they are 'sitting birds'. (There is a Russian satellite in orbit, 'Cosmos 248' in the U.S. classification, equipped for the destruction of satellites. It is believed to have been tested by destroying two U.S.S.R. satellites.) One must wish them further development, because they may be the best safeguard against at least some types of surprise attacks.

* For a terrifying list of these see *Unless Peace Comes*, edited by Nigel Calder, Allen Lane, The Penguin Press, London, 1968.

Biological innovations

3

I will class here not only those innovations which one usually connects with the science of biology, but also all those which deal with living, rather than inanimate matter. I may be forgiven for the inaccurate expression 'living matter'. It is good enough for a rough classification.

3.1 Food

Though in the last years the overall food production in the under-developed countries has increased somewhat faster than the population, there is no ground for complacency. A succession of two bad harvests could easily lead to death by starvation of millions, perhaps of hundreds of millions, and it is doubtful whether the now rather depleted granaries of the U.S.A., Canada and other surplus-producing countries could save them. It has been estimated that by 1975 the United States will have no surplus food for export. Besides, even a small improvement in nutrition would not be satisfactory in a world in which at least one-third, according to others one-half of the population suffers from a monotonous diet, poor in proteins and insufficient to give them the energy required for improving their agriculture, and constructing the huge fertilizer factories, irrigation works, etc. which this entails.

Even without the innovations listed below, existing knowledge joined to reasonable population control would be sufficient to ban not only starvation but also undernourishment from all the world—if only this knowledge plus the will to use it could be brought to the lethargic millions of India, Latin America, and other countries. Colin Clark has remarked that if the Indian farmer could be made to cultivate his land as well as the South Italian peasant, there would be no danger of starvation in India. Another illuminating remark comes from the Hudson Institute (Vol. 4, 'The Year 2000' Context 1969).

The level of rice yield per acre in India at present is that which the Japanese surpassed in the twelfth century.

74. *Improved crops*

Three modern innovations are particularly notable: hybrid maize, dwarf wheat, and a new strain of rice with a yield four times the previous best—though with more fertilizer. By cultivating hybrid maize and dwarf wheat Mexico was able (at least for some years) to increase its food production faster than its record 3 per cent annual population growth. Until fairly recently, the situation of Egypt, with its fast-increasing population, appeared fairly hopeless, because its rice yield per acre was already near the highest in the world, and it was estimated that by 1975, when the Aswan Dam will be full, the new area will be only just sufficient to feed the increased population.

A particularly important innovation would be cereals which could grow in brackish water, because water with a reduced salt content (perhaps 0·8–1·0 per cent) is much cheaper to produce from sea water than pure water. Another way of saving water in tropical countries is hydroponics, reducing the surface through which water can evaporate, and covering the water surface with a layer that reduces the vapour pressure. Unfortunately all of these require more labour, and breaking with long-established traditions.

75. *Improved animals*

The broiler chicken was until recently the paradigm of a cheap and nasty improvement in technology. It ate only about 1·6 times its weight before it could be eaten, but it was a chicken only in name, until it was discovered that it needed only a small addition of chick-weed to its diet to make it taste like a farmyard bird.

The Sacred Cows of India present another type of problem. Fritz Baade, a distinguished nutrition expert, has told the Indian Government that if one-third of them could be killed, the remaining two-thirds would no longer be walking skeletons, and would yield a higher total of milk. The Goverment was willing, but it was almost brought to a fall over the issue. On the other hand the Kobe beef in Japan, reputed to be the best in the world, and the improved breed of cattle in Cuba (milk production doubled in a few years) are examples of what can be done where there is no psychological resistance against innovations.

76. *Fishing and fish farming*

Georg Börgstroem, (*The Hungry Planet*, Macmillan, 1965) one of the world's food experts, has concluded that though fisheries can be greatly expanded, they will not be sufficient to cover the protein deficiency of the world by A.D. 2000 and that the situation in Latin America will be particularly critical. Ocean farming may help, also the cultivation of fish in inland lakes, and in artificial lakes.

It is well known that in the United States pollution has almost ruined the fisheries in the Great Lakes, which threaten to change into an open cesspool. In the first place it was the irruption of the sea lamprey from the Lawrence River which decimated the fisheries, now it is the alewife (*Alosa pseudoharangus Wilson*), which is not yet under control. Restoring the equilibrium of nature and restoring the fisheries at least to their 1945 level will be a long and costly undertaking. $1·4 billion have been mentioned as the cost of controlling municipal and industrial contamination of the Great Lakes over the next twenty years—and this will be sufficient only for preventing matters from becoming worse. The beginnings of industrialization might easily result in similar havoc in the developing countries, and reduce their food supply instead of increasing it.

(IFF: ocean to supply 20 per cent of world's calories, 1985–1992–2000, experts 1992; fish farming to supply 10 per cent, 1990–2010–never, experts 2000.)

77. *Artificial foods*

There are chemists who believe that before the end of the century it will be possible to produce nutrients by synthesis, on an industrial scale. This may be doubtful, but synthesis by micro-organisms, using the energy of sunlight, is already a reality. (Growing yeast on oil, and feeding it to chickens.) For an imaginative look into the future (up to the year 2050) see Nigel Calder, *The Environmental Game*, Panther Books, London, 1967.

(IFF: laboratory creation of proteins for food by *in vitro* cellular processes, 1977–1982, experts 1980.)

78. *Non-fattening foods*

At first sight this might look non-ethical, at a time when at least a third of the world's population is underfed. But this is the invention with the best-assured market, because in the rich countries millions of people are weight watchers. Some may consider it as unethical to pander to the deadly sin of 'gluttony' but it is not unethical in any rational sense, as non-fattening food need not rob the poor countries of their calories.

It is claimed that the carbohydrates of seaweed, which the Japanese consume in large quantities as a 'filler', are not digested by the human organism. Seaweed structured and flavoured to taste like steak might be one of the first of these profitable inventions.

79. *Desalination of sea water*

This has already been mentioned under item 6. Three methods are under active development, distillation, freezing, and selective filtering, with or without electrolysis. (Ion exchange is too expensive.) It appears that this is now more an economic than a technological problem. Water for agricultural use becomes marginally profitable at 20 cents/m^3, fully economic at about 5 cents/m^3. Large distillation or freezing plants can be constructed with such almost complete recuperation of heat that the cost of power need not exceed these figures, but the invested capital is so large that they would pay only at an interest rate of about 2 per cent per annum. (Or a difference between rate of interest and inflation rate of 2 per cent per annum.) This is likely to restrict the building of large desalination plants for the next few years, but in the long run water will be so badly needed that such plants will have to be built almost without consideration of price. The problem may be eased by the breeding of plants that can grow in brackish water. Durable filters are already available which reduce the salt content to 2 per cent.

(IFF: 10 cents/1000 gallons, 1973–1980–1985, experts 1980.)

3.2 Bio-engineering

Medical and biological innovations in the next decades are likely to interfere so powerfully with social life that the term 'bio-engineering' appears appropriate. In medicine and biology there is no clear division between 'pure' and 'applied' science. Louis Pasteur, whose heroic life was a succession of biological inventions, has rightly protested against any such distinction. He would have even more right to protest today, when almost any biological discovery of any importance raises social and ethical problems.

80. *The pill and other methods of contraception*

Far and away the most important and urgent of all innovations. The population explosion does not only threaten the underdeveloped countries with starvation; the comparatively mild increase in the rich countries threatens them equally urgently with frustration.

Several varieties of the pill have been now tested for some years, with undoubted success as regards the reduction of fertility. At the time of writing there is a noticeable rise of fear, amounting almost to panic, owing to the alleged side-effects of the pill. If one remembers the desperate resistance of the British gynaecologists against the mild extension of the abortion law on 'ethical' grounds, one cannot help being suspicious against the doctors who are now loudest in this agitation. There is as yet no sufficient statistical evidence to back them up. One cannot, of course, *a priori* exclude the possibility that the perfect pill has not yet been found, but one can leave this safely to medical research backed by the new, fast methods of computerized medical statistics.

Propagating family planning belongs more properly to social innovations than to bio-engineering.

(IFF: economical mass-administered contraceptives 1972–1983, experts 1983.)

81. *Cancer cures*

If a popular vote were taken, a cure for cancer would probably top the list of desirable medical discoveries. It is generally agreed, for a long time, that this is not a problem for the medical profession, but for fundamental research into the processes of growth and reproduction. (This does not of course mean that it belongs to 'pure' research; any such research becomes applied at the moment when it is successful and ceases to be 'idle'; Whewell's term for what is now called 'pure' research.) The wonderful progress of molecular biology since the breaking of the genetic code has undoubtedly encouraged the optimistic forecasts of experts regarding chemo-therapeutic cures for various types of cancer, which I quote below without comment.

(IFF: 1983–2000–2015, experts 1990.),

82. *Theoretical pharmacology—prediction of the effects of drugs*

Pharmacology is only half a science; the chemical composition of the drugs is accurately known, but their effects can be found out only in animal experiments, and often come as complete surprises. Prediction of these effects will be possible of course only with the complete understanding of the biochemical processes, but progress is so rapid that fairly confident forecasts can be made.

(IFF: 1990–2010, experts 2000.)

83. *Ultrasonics instead of X-rays*

The application of X-ray diagnostics in cases in which they could do hereditary damage has been greatly restricted (no X-rays in ordinary pregnancies, only one in difficult cases), but heredicists are still far from happy. On the other hand ultrasonic waves have no chemical effects (except at intensities so extreme that they produce cavitation and disruption of tissues, or by prolonged heating), their penetration is sufficient, and at 10 megahertz or higher their wavelength is short enough for the required detail. They are particularly suitable for pregnancies, because of the absence of air cavities, and impressive results have been already obtained by echo (sonar) methods, which produce sections of the body piecemeal. Imaging methods were so far not very successful, but they are highly desirable and can be confidently expected to succeed in the near future.

84. *Immunizing agents which protect against most bacterial and viral diseases*

The *ars sterilisans magna* of the old doctors. Antibiotics with very wide spectra are now in existence, but immunizing agents act only in specific diseases. But as there are individuals with natural immunity against almost any disease, it does not appear impossible to fight a wide variety of bacteria and viruses by artificially enhanced immunity. This probably explains the surprisingly optimistic forecast of experts, quoted below.

(IFF: 1973–1980–1985, experts 1980.)

85. *Transplants—fighting the foreign body rejection of the organism*

Transplants of hearts and other organs are the most spectacular results of modern surgery, but their success has been limited so far by the weakening of the organism caused by the anti-immunizing agencies which were required to prevent rejection. So far nobody with a transplanted heart has lived for more than eighteen months, in most cases the lease of life was much shorter. Much effort is now going into the search for agents which prevent rejection without

de-immunization against viruses, and immunizing agents which do not cause rejection. The forecasts are very optimistic.
(IFF: 1973–1982, experts 1983.)

As organs from an identical twin who has died just at the right time are hardly ever available (and would be probably useless if they were), some biologists have thought of breeding tissue-compatible animals, whose organs would serve as well as (or better than) those of identical twins. This would of course require an enormous extension of the objective knowledge of compatibility, and the prospect is fairly remote.
(IFF: 1990–2015, experts 2015.)

86. *Artificial hearts*

Transplanting hearts or other organs from fresh corpses is not a solution which can satisfy anybody in the long run. Artificial hearts, i.e. blood pumps outside the body, have been tried out with some measure of success in animal experiments. Artificial hearts inside the body, which will not be rejected and which contain a long-duration power source, constitute an enormously difficult problem, but experts are remarkably confident.
(IFF: 1980–1990, experts 1980.)

87. *Renewal of organs*

Renewal of organs, such as of a worn-out, old heart, by stimulating natural renewal processes is a far more attractive prospect than any implanted machine. It presupposes a colossal extension of our knowledge of the factors which control organic growth and ageing. Some experts are inclined to believe that it may be achieved, or at least enhanced, by establishing in the old organism the hormonal levels of a young one.
(IFF: 1980–2010–2025, experts 2010.)

88. *Repair of neurons*

There are some encouraging recent findings which indicate that severed neurons can grow together in the right ambient. (In a living artery.) Whether neurons can be induced to grow into an old limb, in which they have died off, as they grow in an embryonic animal, remains highly hypothetical. The matter of 'brain transplants', which has been boldly mentioned as a possibility, is hypothetical to an even higher degree.
(IFF: 1990–2010, experts 2000. Laboratory demonstration only.)

89. *Ageing postponed*

Geriatrics is now an important branch of medical research. It cannot be said so far that it has extended human age beyond its 'natural' limits, but it has certainly enabled 'medicated survival' of many, and a reasonably complaint-free old age for even more, especially for women past the climactery. In view of the population explosion, an extension of human age by fifty years, as contemplated by the forecasters quoted below, would be a very doubtful blessing at the present epoch, and may be more desirable perhaps a hundred years hence when humanity will be, one may hope, 'mature enough for maturity'.

(IFF: 1990–2015, experts 2015.)

90. *Early detection of abnormalities in the foetus by chromosome tests*

Such a test is already in existence. A sample of the placental fluid is taken with a syringe. There are always a few cells of the foetus floating in it. A culture of these is made (on beans) and the chromosomes of the new cells are spread out on the surface of a special fluid agent. It is claimed that the extra chromosomes of mongols and of other abnormalities can be detected with great certainty, well before the third month when abortion is easiest. In principle, if such tests could be made compulsory, no more mongols need be born. (At present the total fraction of abnormals is about one in two hundred, and causes an immeasurable amount of unhappiness.) In view of these claims put forward by highly respected researchers I cannot understand why the experts consulted by the IFF consider such tests as yet to be discovered.

(IFF: 1980–1990, experts 1980.)

91. *Preselection of the sex of babies*

This is a paradigm of scientific progress which adds one degree of freedom to those of humanity, and yet might restrict subjective freedom to a painful degree. In principle the above quoted chromosome test is one way of preselection, because the sex of the foetus can be determined from the chromosome, and the undesired sex could be aborted. Pre-determination of the sex by fractioning sperms had some success in the artificial insemination of cattle. Assuming now that some such method were sufficiently acceptable to be practised on a grand scale, it raises immediately a social question. Our whole social life (and not only our sex habits) is founded on the approximate equality of numbers of the sexes. If a survey were made, and it were found that the majority of young parents-to-be are in favour of the English 'pigeon-pair', a boy first and then a girl, everything would be well and good. But if it turned out that there is a serious imbalance in the wishes, it would be much wiser for legislation to forbid the whole practice than to issue 'boy' and 'girl' permits.

(IFF: preselection with 90% certainty, 1980–1990, experts 1980.)

92. *Fertilization of human ovum* in vitro, *implant in host mother*

Fertilization *in vitro* has been recently achieved with a human ovum, implantation of a fertilized egg into a host mother was successful in rabbits and other animals. Applied to humans, there is a possibility of 'forced eugenics', rather more complicated than artificial insemination from a stock of selected sperms, but even more selective, because the ovum could be selected too.

(IFF: 1980–1990–2010, experts 1990.)

93. *Human clone*

This is a prospect a whole degree more hypothetical than the last. It means that the nucleus in a human ovum is replaced by a somatic cell of a human being, and introduced into the placenta of a host mother, where it develops into an identical twin of the first. (Rather less efficient than Aldous Huxley's 'bokanovskified twins', in batches of, say fifty, to serve fifty identical machines.) It is hard to see why one should need twins of outstanding creative individuals when their output will be the same?

(IFF: 1990–2000–2010, experts (N.B.) 1985.)

94. *Extra-uterine development*

This is the old '*Ectogenesis*' of J. B. S. Haldane (1922) and of Aldous Huxley (1931); human embryos brought up in a simulated placental environment, in a 'hatchery'. At the time when Haldane wrote of it he had serious doubts whether human beings had sufficient interest in breeding in the natural way for maintaining their numbers. Haldane, like all of his contemporaries, entirely failed to foresee the population explosion. As this explosion is now with us, ectogenesis is a rather bizarre prospect, but it is interesting to see that biologists do not seem to consider it any longer as fanciful nonsense, but as a somewhat remote possibility.

(IFF: 1990–2015, experts 2015.)

95. *Chemical control of heredity, by molecular engineering of genes*

This is probably the boldest of all projects in bio-engineering, and it is remarkable that it appears to have been first proposed (in 1961) by Joshua Lederberg, who is one of the loudest protesters against more orthodox methods of eugenics. Lederberg's proposals were attacked with at least equal vehemence by the physicist Walter Heitler (*Die Manipulierbarkeit des Menschen*, Mainzer Universitäts-gespräche, Sommer 1968) and by many others.

The idea is to influence, by chemical substitution methods, the heredity of man as encoded in his DNA-chains, with a total length of about two metres. *If* it were effective, it would be undoubtedly incomparably more efficient than selective, eugenic breeding. On the other hand all the doubts which have been brought up against eugenics by Haldane, Medawar, and many others would apply to it *a fortiori*. Man would take his phylogenesis into his own hands. The most pregnant comment on this was made by Jean Rostand: '*Où apprendre le métier de Dieu?*' (Who can teach us God's business?)

(IFF: possibility, not necessarily acceptance, 1990–2000–2015, experts 2020.)

96. *Drugs to improve perception and learning speed of retardates*

Certain drugs, such as benzedrine or amphetamine, taken in moderation, can temporarily improve the performance of normal individuals, by helping to concentrate and by excluding diversions. They have almost the opposite effect when taken in excess. It appears likely that unless they prove to be habit-forming and lead to excesses, they will improve retardates too, but the issue is doubtful.

(IFF: 'LSD-like drugs', a somewhat unfortunate formulation of the question in view of the disastrous record of LSD, 1975–1990–2010, experts 1985.)

97. *Aldous Huxley's 'Soma'*

A drug which gives a happy holiday from the real world, but unlike alcohol does not lead to quarrelsomeness or violence, and does not leave a thick head. Aldous Huxley believed, and I do not think that Siegmund Freud would have cared to contradict him, that no high civilization is possible without an easy escape from reality. Huxley was only grievously mistaken when he thought that mescalin, LSD, or psylocybin are acceptable substitutes for Soma.

98. *Drugs that permanently raise the intelligence*

No such drug exists at present. If it were found, it is likely that it would have to be administered in the very early years, perhaps even prenatally. There are claims that pre-natal oxygenation and/or relief of pressure on the abdomen of the mother can produce babies of superior intelligence, and if this can be proved, it may be a cue.

(IFF: 1980–2000–2020, experts 2020.)

99. *Drugs that change the personality*

Such a drug, by no means impossible, is right on the borderline to a social invention of the greatest importance, because it is fraught with social consequences. Arthur Koestler, in his book *The Ghost in the Machine*, 1967, has argued very impressively that Man is one of the many mistakes of the creation. His mighty neocortex is only the slave and executor of the inner brain, which is still much the same as that of the ancestral, carnivorous, and bloodthirsty ape. Man has a fatal weakness not so much for personal aggression, as for collective madnesses; witness our sad history of wars and of human sacrifices. Man will certainly destroy himself, unless his nature is altered forcibly by some drug which could be added to the drinking water or to salt, to suppress his aggressive instincts, and his urge towards collective sacrifices.

Drugs which pacify individuals are already in existence. *Diazopan* is reported to turn the wickedest boss-monkey into a pacifist. Whether it can also suppress the urge for identification 'with a cause to die for' is not known, because this weakness is specifically human; monkeys are less subject to it.*

There are of course many objections against trying such drugs on humans, especially on a grand scale. Will they not also suppress willpower and creativity? Could they not be used for the submission of other nations, for the pacification of minorities with just grievances,

---

* To be fair to monkeys, some of them seem to have advanced to the stage of tribal patriotism. Robert Ardey in his *African Genesis* reports an astonishing observation of the South African amateur zoologist Eugene Marais. One night he observed a herd of baboons, and saw a leopard on a rocky ledge above them, preparing to pounce. Then he saw two male baboons climbing quietly above the leopard, and throwing themselves on it at the same moment. They were dead in an instant, but the leopard was also dead; his jugular vein was bitten through by one of the heroic baboons.

even for establishing the rule of an authoritarian minority over a spineless, submissive majority? It is most unlikely that humanity will adopt it universally before it is stricken by a catastrophe so terrifying as to make it profoundly disgusted with its own nature. Until then pacifying drugs may well be used only as weapons; but rather preferable to all other weapons.

(IFF: 1973–1980, experts 1980.)

100. *Creation of some sort of artificial life*

When Wöhler synthesized the first organic compound, urea, in 1828, there were some people who believed that the next step would be the *homunculus*. After 150 years of progress we have become more modest in our aims; we would be satisfied with some sort of virus, or even a self-reproducing molecule. Such a discovery would have a rather exceptional standing in biology; it would be a scientific achievement of the highest order, but for the start at least without great social consequences. (In old times, when religions were interpreted dogmatically, the impact would have been tremendous.) The opinions show a remarkable consensus.

(IFF: 1977–1983, experts 1980.)

Social innovations

It is a commonplace in our days that social development has not kept up with the explosive progress of science and technology. It was probably J. M. Keynes who put the chief reason for this into clear, quantitative terms. Very few people change their view and values after their school days. Something like twenty-five years elapse between the university and the graduate obtaining a position of consequence. If during this time the technological basis of society has significantly changed, the conditions are given for a serious mismatch. The great originator of the quantum theory, Max Planck, has expressed a similar thought: 'Theories win recognition not so much by acceptance, as by the dying off of their opponents'. This is rather too pessimistic, and in fact if the time lag in science and technology were as great as in the institutions, we would have far less of a problem.

The worst part of the problem, which only few people have had the courage to face, is that science and technology may have created a mismatch not only with the institutional structure of our society, but also with the basic nature of man. If we solved our social problems, if we succeeded in ensuring a secure life for every child born, from the cradle to the grave, what is there left to fight for? David Riesman has expressed this in merciless words: 'What we fear most is not total destruction, but total meaninglessness'.

This is not yet a problem for today, but one which will become inceasingly important in the economically successful part of the world in the next thirty years, and we must start thinking about it right now. The problem is not as easy as the well-meaning social reformers of today imagine it, who work, rightly, towards better housing, better medicine, etc., nor is it as hopeless as the prophets of doom believe, or pretend to believe. There have always been some privileged people, whose security was assured from birth, and for whom life nevertheless was full of meaning. Perhaps it will be possible to mould the great majority of people to this pattern, without drugs.

The social innovations listed below are mostly *problems*. In some cases there appear to be solutions on which there is a reasonable consensus, in others I have hinted at possible ways of dealing with them. These ought to be considered less as proposals, more as subjects for discussion. The talents of a whole generation of thinkers will be needed for solving them, and of a whole generation of men of action for putting them into practice.

I thought it better to repeat, in many places, the technical means of improvements, already mentioned under 'hardware' innovations, rather than referring the reader to the previous entry.

4.1 Human ecology and ekistics

Ecology is the science of the interaction of living species, plants, animals, or men, with their environment. Interest in human ecology has enormously increased in the last years by the awareness of the pernicious influence of our civilization on the biological pyramid on which man is living, by insecticides, by the pollution of air and water, and by upsetting the CO_2 equilibrium in the atmosphere. It may give us some hope that in the thirties there was a first wave of ecological panic, caused by the danger of soil erosion. This now appears to be overcome in the U.S. by appropriate legislation. On the other hand, the wounds caused in the ecology of Pakistan by water regulations, which caused thousands of square miles to become unfertile by salination, still remain to be healed.

Ekistics is the term introduced by the eminent Greek architect C. A. Doxiades for the 'problems and science of human settlement'. This can be considered a particularly important sector of human ecology. The unmistakable tendency in all the world at present is *urbanization;* the crowding of an ever-increasing part of the population in towns. If the present trend continues, by the year 2000 at least 75 per cent of the U.S. population will live in large cities. Moreover, this will lead, by that time, to the formation of giant cities, *Megalopolises*, with 50–100 million inhabitants. Herman Kahn foresees three such monsters in the U.S., which he calls, only half in joke, 'Boswash' (from Boston to Washington), 'Chickpitts' (from Chicago to Pittsburgh), and 'Sansan' (from San Francisco to San Diego). In the U.K. there are already two enormous conglomerates, in the London area and in Lancashire, but owing to the slower growth of the population, and the effort of the government to create new towns, neither of these aggregates will grow to a Megalopolis by A.D. 2000. But if we look further ahead, the prospect becomes frightening. By the year 2050 the U.S. expert Richard Meier expects a town around Bombay with 500 million inhabitants, and Doxiades visualizes an 'Ecumenopolis', extending almost without interruption from London to Peking.

These things must not happen. Let us follow Lewis Mumford's advice: 'Do not consider a trend as a command!' The beautiful cities of the past, like Paris or Florence, were flowers of civilization, but Megalopolis is the end of it. Only dull people (perhaps dulled by constantly taking pills) could survive in them, because Man is not likely to develop the social instinct of an ant in one or two generations. Most of the innovations which I list below aim at fighting Megalopolis. For a while this will be of course a losing fight; we cannot expect to turn the dangerous corner so long as the population growth continues, but at least we may be able to ward off its worst consequences: universal frustration, even despair.*

101. *Stopping population growth*

This is the Alpha and the Omega of all efforts for stopping the advent of Megalopolis, but we cannot achieve very much in the short run. The technical means at our disposal; the pill, the intra-uterine loop, abortion. By legalizing abortion Japan, Hungary, Romania, and Czechoslovakia have almost stopped growth; in fact their net reproduction rate has dropped below unity. It will be a long time though before the growth of the developing countries, which is now at about 2·5 per cent per annum (doubling in thirty years) can be expected to stop, and with all efforts at family planning the world population in A.D. 2000 will hardly be less than 6 billion. Even in the United States the excess of births over deaths is 2 million per annum and the population will probably approach 300 million by A.D. 2000. In Britain the growth figure has now dropped to about 250 000 per annum but even this moderate increase presents difficult problems in

* A spirited attack on the horrors and dangers of urbanization is contained in E. J. Mishan's book, *The Costs of Economic Progress*, Staples Press, London, 1967, pp. 74–106. My own views are closest to those of Lewis Mumford, which he has expressed with style and passion in a long series of great books in the course of forty years.

an already highly populated country. Five or six large new towns ought to be built every year to accommodate them. Space would be scarce even without an increase of population, because the workers in a rich, modern industrial society will not be satisfied with the back-to-back hovels of the nineteenth century, of which there are still far too many left in our time.

In the underdeveloped countries, such as India (where the government powerfully supports it), and in Latin America (where it comes up against strong religious resistance), the propagation of the technical means of contraception is the most urgent task, and this applies also to the American ghettoes. But it will not be enough in the rich countries, in which hygiene has developed to such a degree that not more than 2·1–2·3 children can be allowed per couple for a stationary population. In this third of the world nothing less is likely to work than a form of taxation; no allowance after the second child. This is probably the limit to which legislation is likely to go in this century. Kenneth Boulding (in *The Meaning of the 20th Century*, 1963) has proposed a more radical remedy, but without any hope that it will be accepted; *negotiable* breeding permits, one per person. The more responsible people would buy them from the less responsible ones. For the time being we must not expect that the unavoidably urgent reduction of the growth rate can be coupled with eugenic mechanisms such as this; even the 'democratic' reduction will be difficult enough.

102. *Smaller towns, with good interurban transport*

In the antique world, a democratic city could not well extend beyond the reach of a speaker's voice. In our world there is no such limitation (think of the loudspeakers that carried Goebbels' voice over the whole of Germany), but we have good reasons for considering 50 000 as a reasonable number of people in a modern town. Doxiades wants to build up his giant towns of such units, suitably separated, not merged into one big rabbit warren. There is room for these, even in highly populated countries. A city of this size can have all the amenities, its own sports stadium, and its own live theatres. (Like the Greek cities, or the German residential towns.) There would be no need for intense commuter traffic between them, because each town could offer jobs for its own population. But so long as the architect is bound to provide parking spaces all over the town, these small cities could hardly become things of beauty, like the historic ones. There seems to be now fairly general agreement among town planners that no cars, and not even bicycles, ought to be allowed near the town centre. These ought to have plazas, for pedestrians only, with cafés (such as now hardly exist except in the San Marco of Venice), where people could walk, talk, and enjoy their leisure. Garages (except underground ones) ought to be allowed only near the periphery. Fast electric trains ought to connect these towns with one another, and with the airports.

When spinning out such modest little utopias, one comes up every time against the question: 'Who shall pay for it?' Since the advent of the motor car, many railway lines have had to be shut down in England because they were not profitable, and in the States quite a few vitally important railway lines are kept grudgingly going— under receivership. The motor car is still gaining on other means of transport, except by air. Many Americans believe that the U.S. economy is based on Detroit turning out 9 million cars per year, with almost as many going on the scrapheap, and they are right in the short run—in the very shortsighted run. Evidently the law of supply and demand cannot be expected to create countries very different from those of today; giant cities with sprawling suburbs. Only enlightened legislation can help.

103. *Decentralized production*

There are few industries today in which the optimum size of a factory is not at or below 5000 workers, and in most industries optimum efficiency can be reached at 1000 workers or even less. Nevertheless the tendency of large works to settle in large conglomerations is still as strong as ever, because it is in such areas that labour can most easily be found. It is only counteracted to some extent by the difficulty for the workers of finding inexpensive housing, and the result is an unhealthy, exaggerated commuter traffic. This is enhanced by economic uncertainty; not all workers are willing to rely for their livelihood on the one local factory. (Japan, where the workers are wedded to their firm, is an exception, but even Japan has not yet taken advantage of this important factor in favour of decentralization.)

So long as firms can shut down factories from one day to another because they go bankrupt, or are forced by competition to rationalize and to merge manufacturing units, or because labour difficulties make them start afresh with a new factory in another place, no great improvement can be expected. Labour will have to remain mobile, settling permanently only within the area of big cities, and this geographical mobility (unlike 'social mobility') is an enemy of any sort of high, stable civilization. In the epoch of rapid change and economic uncertainty that is still before us, the initiative has to come from enlightened employers, supported by their governments. (The decentralization of General Electric under the Chairmanship of Ralph Cordiner is a fine example.) Modern technology is on the side of the decentralizers, by creating all the possibilities of communication and transport between the centre, and between the manufacturing units.

104. *Communication between decentralized units*

The technical means have been discussed under No. 39. They are all available or, as in the case of a drawing which has to be discussed at both ends though it is physically present only at one, they are matters of straightforward development. As in most social innovations, the problem is one of economy, not of technology. For instance telephone lines used to be allocated up to now on a strictly commercial basis, so that the queuing would just be tolerable. This has worked out quite well in cities, but more than one London firm which, following the advice of the Government, had moved its seat to neighbouring towns had to return to London, because the telephone lines were always engaged. It appears that decentralization cannot start in earnest before telephone lines, and also telex lines, even coaxial cables and helical wave guides, get subsidies as social services, instead of having to pay for themselves.

105. *Teleshopping*

Edward Bellamy in his *Looking Backward* (1888) was the prophet of shopping in sample-stores, with credit cards, with the goods being sent home by post from large warehouses outside the city. This, like the broadcasting of music, was one of the dreams of this highly practical utopian which have been at least partly realized, by Sears-Roebuck and others. Nevertheless shopping streets, like Oxford Street in London, still constitute some of the worst bottlenecks for traffic. The development towards the postal-order store has gone into reverse with the spreading of the motor car, which made it possible for housewives to take home far more heavy and voluminous shopping than they would have been able to carry.

Modern means of communication could enable us to go even beyond Bellamy. The housewife could make her choice from an

EVR film or video tape on her colour television set, and order it by dialling on her telephone. The goods could be delivered to the home, as Mishan has suggested, by silent electric cars between 3 and 7 a.m. in the same way as newspapers and parcels are distributed in suburban or rural districts in the States. Of course what the housewife would do with her spare time and energy is another problem.

106. *Decentralization of entertainment*

In this century the cinema, radio, and television have gone a very long way towards demolishing the privilege of the large city in entertainment. Even international sports events can be enjoyed on the television screens almost as much as by being present, without the discomforts which are unavoidable when 100 000 spectators are crowded together. Nevertheless, they have not perceptibly reduced the mass migrations to great sports events, nor the popularity of the live theatre. (Except the music halls, and the amateur theatricals in the North Midlands, of which J. B. Priestley writes with such nostalgia.) It would be a great exaggeration to suppose that the live theatre plays any great part in the mass-migration towards the great cities, but I believe that a considerable attraction could be added to the small new towns if they had their own theatres, and not only their cinema palaces and bingo-halls.

The following four items deal with a problem of human ecology which is now very much in the centre of the stage: pollution. In all cases we have the technological means to fight them, and the question is only: 'who shall pay for them?'. It may be noted that the economic problem involved in pollution is very much smaller than that of urbanization. Rebuilding Britain on the lines visualized by the new town planners will absorb a great part of the whole national effort for one or two generations. On the other hand the highest estimates of the cost of fighting pollution amount to not much more than $1\frac{1}{2}$ per cent of the G.N.P.—and yet it is a very serious problem if it has to come out of the pocket of the taxpayer.

107. *Air pollution*

Modern factories, even steel works, are no longer the 'dark satanic mills' of Blake, and they no longer drop 1–200 tons of soot per year on a square mile of London or Leeds. They also use oil in increasing proportion, with a sulphur content, which, though not negligible, is smaller than that of lower-grade coal. On the other hand air pollution by motor cars has increased enormously. Contrary to popular belief, jet planes contribute only very little to it.

The most radical method of reducing air pollution is outlawing the internal combustion engine, and at the time of writing there is a popular demand to this effect circulating in California, which is reputed to have collected more than 100 000 signatures. It is always doubtful whether extremist demands by minorities can have far-reaching social effects, but they can serve as social indicators, like the suicide figures for general unhappiness. The next most radical method is the exclusion of petrol-burning motor cars from city areas, and replacing these by electric cars. As previously mentioned, as electric cars based on fuel cells or new types of primary batteries now seem to offer little promise of a speedy solution, these cars ought to be powered by light lead accumulators, interchangeable at what used to be filling stations. Less radical methods aim at retaining the internal combustion engine, but reducing the unburnt hydrocarbons and noxious combustion products. There appear to be three ways towards this. The first is an after-burner, to ensure complete combustion of the exhaust gases, and is not very promising. The second is computer-controlled injection of fuel, ensuring optimum combustion at every speed and during acceleration. This is fully developed, and held back only by the extra cost. The third is an improvement of the fuel, such that it ensures good anti-knock properties, without lead additives. A particularly interesting variant of this is the removal of the long-chain components in the fuel, which are chiefly responsible for the knocking, by molecular sieves. These could then be used for cultivating protein-producing micro-organisms, such as yeast, which grow particularly well on the long-chain components. Non-lead additives are also likely to make their appearance. All of these methods would add to the cost of the motor car and/or of the fuel, and will not gain acceptance unless enforced by legislation. It would be too much to ask the individual motorist to spend more on his car, so long as he is surrounded by hundred thousands of motorists who have done nothing against pollution.

(IFF: 1980–1990, experts 1980.)

108. *Water pollution*

In 1969 chemicals from leaky containers in a barge killed all fish in the Rhine. In the United States the Great Lakes have perhaps ten years to live before they become cesspools. The *Torrey Canyon* and the underwater burst of oilwells near Santa Barbara have produced spectacular examples of what man can do even to the sea. In all these cases the damages were far in excess of the costs of their prevention.

Legislation is now being prepared in most industrial countries for the control of the effluvia of chemical factories, and the U.S. budget has $900 million allocated for the fight against water pollution. Action is being taken also against another type of pollution; by plastic containers. (Thor Heyerdal has found them swimming even in the Pacific, and the Lake of Geneva is choked by them.) In Germany the use of all non-biodegradable plastic containers has been outlawed, and other countries are likely to follow suit. This, at any rate, is a type of pollution which is likely to be overcome in the next decade.

(IFF: efficient disposal of solid waste products, 1980–1990, experts 1980.)

109. *Insecticide pollution*

We are all indebted to Rachel Carson, who in *Silent Spring* (1962) first raised her powerful voice against the indiscriminate use of insecticides. DDT, a powerful neurotoxic, was not so long ago the best weapon against malaria, sleeping sickness, and other insect-born diseases, but in recent years it has not only killed off singing birds, but is threatening also humans. It has been found even in penguins. It is not yet clear what will take its place. The value of DDT, like almost any agency used for a long time, has diminished anyway by the selective breeding of DDT-resisting strains of insects. On this front, as on many others, nature has hit back, and constant watchfulness will be necessary. Watchfulness is also needed against the indiscriminate use of much more powerful agents: 'In March 1968, grazing flocks of sheep were victims of the first widely publicized CBW accident when an Army plane released 320 gallons of VX, a long-lasting nerotoxic, anticholesterase, high over its Dugway, Utah proving grounds…When a sudden wind shift carried the gas from the proving grounds to nearby grazing spots, 6400 sheep were killed.' (*The Sciences*, New York Acad. Sci., **9**, 9 Sept. 1969.) If the wind had carried in another direction, it might have imperilled the lives of the 200 000 inhabitants of Salt Lake City. It may be hoped that at least such accidents will not be repeated after President Nixon's decision to stop all Chemical and Biological Warfare research, and to destroy the stockpiles.

110. *Noise pollution*

Among other benefits, the electric motor car would greatly lower the noise level of cities. On the highways, where the internal combustion engine motor car is likely to stay for a long time, motorists often indulge in making unnecessary noise. A device which takes care of this has been developed in CBS Laboratories, Stamford. It measures the noise level of the car, and if it exceeds a certain limit, it takes a photograph of its licence plate. Like the radar speed trap, the knowledge or suspicion of being watched has a highly salutary effect on motorists.

Unfortunately, little can be done against the noise of jet planes. Any reduction of noise goes at the expense of efficiency,* though some progress has been achieved with the Boeing 747 engines and the Rolls Royce RB 211. The best remedy is to move the airports far away from towns, and to forbid aircraft to fly low over them. Whether this will be sufficient to prevent the dreaded effects of the supersonic boom remains to be seen.

4.2 Fighting crime and corruption

According to Scotland Yard in England, and the F.B.I. in the United States, crime is increasing in these two leading countries roughly at the rate of 10 per cent per annum. These figures have been often doubted by well-meaning liberals, and attributed to better methods of crime detection, or to the desire of the police authorities to increase their establishment. However, they are now well backed by the insurance companies, some of which have got into serious difficulties. In addition to crimes committed for material gain, there seems to exist an even more frightening increase of crimes 'for kicks'. In Britain, between 1939 and 1961, crimes of violence by the 14–21 age group have increased nearly fifteenfold. (T. R. Fyvel, *The Insecure Offenders*, London, 1961.) The cry for 'law and order' is now very loud, both in Britain and in the States. In the United States, in November 1969 a committee headed by Milton Eisenhower published a remarkable report. Unlike so many social surveys which avoid conclusions, and usually end with words to the effect that 'with more money and a more comprehensive survey some definite conclusions may perhaps be established', this report boldly extrapolates the present trends into the future. *If* crime continues to increase, the

---

* Ariel Alexandre, in a study on aircraft noise (Centre d'Etudes et de Recherches d'Anthropologie Appliquée, 1970), states that reducing the jet noise from 130 dB to 110 dB would raise the cost per ton-mile by 70 per cent, a reduction to 105 dB by 140 per cent.

rich people will concentrate in areas well guarded by their own, heavily armed police. These areas will be connected by roads, also heavily guarded, where the rich people will be able to travel in comparative safety in bullet-proof cars. The slums will be left to become festering morasses of crime. Such 'if' pictures are not of course forecasts; their usefulness is in being self-annulling predictions.

111. *Crime prediction*

There is no difficulty in statistical crime prediction. The police know with fair certainty how many crimes can be expected in the next year in a given district. *Individual* crime prediction is only in its infancy, but it shows great promise. A particularly encouraging start was made by William and Joan McCord (*Origins of Alcoholism*, Tavistock Publications, London, 1960) in a survey of a strongly alcohol and crime-ridden district of Massachusetts. They concluded that given certain family conditions, it could be predicted with some 90 per cent probability that the child will become an alcoholic or fall foul of the law before he reaches forty. It is possible that the McCords somewhat overemphasized the influence of environment; some of their slum-dwellers may have been innately predisposed to crime. (For instance those who had an extra chromosome, but this is found only in 2 per cent of the criminals.) However this may be, the social reformer must take notice of the facts of crime-breeding environment, and take measures for combating them, however difficult this may be. The most radical method would be of course the dispersal of the ghettoes, and taking away the children from unworthy parents, to be brought up by trustworthy foster-parents. The size of this task may be judged from the fact that in Britain almost 40 000 cases of cruelty to children, mostly committed by parents, come into court every year, and the children are taken away from the parents only in a small fraction of these cases. Liberal conscience rebels against making this practice more general, and extending it also to cases where there is only a suspicion of danger to the society, but it appears that if we consider the right of parents to bring up their children as they like as an inalienable freedom, we may as well resign ourselves to a crime- and alcohol-ridden society.

112. *Slum clearance*

This at least is a programme about which everybody is agreed in principle, opinions differ greatly only regarding the extent to which this is to be paid for by the taxpayer. The task is enormous, in the United States it has been estimated as of the order of 20 billion dollars per annum for at least one decade. There is one observation which may be helpful. Human nature is such that it does not appreciate what it gets for nothing. It is not surprising that it has been observed in the United States that when the inhabitants of ghettoes were moved to ready-built suburbs, they soon turned these into slums. On the other hand where they had to take part in the building, or to bear part of the costs of the new habitations, the results were much better. Consistently applied, this principle may be of some help in speeding up slum clearance without excessive costs to the taxpayer.

113. *Crime prevention by observation*

This is a subject which one approaches with very mixed feelings. Using informers for warning the police of crimes, assassinations, and riots are the hateful methods of the police state. Tapping telephones is a little less hateful; it does not involve the degradation of a human being. As Nigel Calder remarks, who could object to it if it is a question of saving a kidnapped child? But to be effective, the net must be thrown so wide that it must lay bare the private lives of many more innocent people than criminals. Bugging hotels with microphones is easy; bugging the slums is almost impossible. Anybody who has read Solzhenitzin's great novel, *The First Circle*, 1968, must have a violent revulsion also against the tapping of public call boxes. Thus the scope of crime prevention by scientific observation is rather limited. Nobody can object to the surveillance of streets by television cameras, to prevent thefts from cars or of cars. It would be also desirable to identify cars not with licence plates, which can be easily changed, but by marks irremovably engraved in the body.

114. *Large-scale, organized crime*

The wholesale profit on the hard killer-drug heroin in the United States is now estimated to amount to 300 million dollars per annum. Crime on such a scale is hardly possible without corruption. The chain between the poppy-growers in Anatolia and the unfortunate customers is such a long one that it ought to be possible to break it at many points if the international agreements were enforced. It may be easiest to break it at the source: the total value of the Turkish poppy harvest is less than a million dollars. The same applies to LSD, because though almost anybody can make LSD out of lysergic acid, the manufacture of lysergic acid requires expensive equipment and expert chemists. Unfortunately, the development of mass tourism is making the job of the police more and more difficult. The problem is such a hard one that even some liberal-minded people could think of nothing better than the electric chair for drug peddlers caught the second time. I do not wish to propose any special solution but recommend the problem to the attention of inventive-minded people who would like to do a great service to society.

115. *The traffic in stolen goods*

A certain amount of pilfering is tacitly tolerated. For instance the English dockers considered it as their traditional right to retain one tea chest in a hundred for their own use. There can be little objection to this, but it is alleged that it is one of the reasons for the strong resistance of the English dockers to containerization. There must be a better way for satisfying them than, as it were, paying a part of their wages *via* the insurance companies. Observing the dockers and other transport workers by invisible television cameras would probably meet with violent resistance. It may be a better way to mark all goods in such a way that they can be traced to the small traders and to the markets.

116. *Elimination of cash payments*

This is well on the way, especially in the United States, but wages are still mostly paid in cash, and the robbing of pay-waggons and of banks is still an almost daily occurrence. There are still many shops and even hotels where cheques are not accepted. Electronic verification of cheques, also vouchers and credit cards carrying a photograph, or, even better, a holographic photograph of the rightful bearer, have been proposed and are likely to produce a great reduction in cash payments.

Once the stealing or robbing of cash has been reduced, the robbing of goods is likely to increase, and the marking of goods, as suggested in the previous entry, will become even more important.

117. *Crime 'for kicks'*

This is not a matter for law enforcement by economic and techno-logical innovations, but for prevention. So long as there are ghettoes, there will be gangs of juvenile delinquents. The problem is more or less what William James (1905) has called 'The Moral Equivalent of War'. We will come back to it later.

4.3 Monetary and economic reforms*

A modern democratic state should endeavour to satisfy four conditions simultaneously:

Balance of foreign payments
Full employment
Steady growth
Stable prices.

In the last decade not even Germany, that roaring economic success, has managed to satisfy completely all the four conditions. Prices in Germany have risen at a mean rate of 2·5 per cent per annum over ten years, but in 1969–70 at 3·8 per cent. Similarly, Japan, the country with the steepest economic growth, and Sweden, the model country of social peace, had and have permanent inflation. Britain and the United States had to slow down economic growth, and yet could not stop prices increasing at the rate of 5–6 per cent per annum.

There are two chief reasons for this. One is that inflation produces favourable conditions for economic growth, rising prices promise good profits and inflation allows to pay back good money with bad money. On the other hand inflation means the permanent exploitation of old people by the young, and social insurance is a very doubtful remedy. The other reason is the 'revolution of rising expectations'. In the industrial countries the workers expect to be better off next year, and their demands overshoot the rise in production. This makes inflation a self-catalytic process; the expectation of inflation produces inflation. Trade unions which notice that their gains in wages have dropped, say, 8 per cent below the price index, demand a 20 per cent rise to be safe in the next years.

The Soviet Union also had to fight inflation at various times, but communist countries on the whole have managed to cope with this problem, though they could not quite dispense with the safety valve of a tolerated 'free' market. The wages are calculated so that the workers shall be able to consume what they produce in consumer goods, plus a little in savings towards a house, or retained in the hope of

* In this section I had the assistance of my brother André Gabor, of the Department of Economics in the University of Nottingham, which I much appreciate.

being able to buy goods next year which are not at present available. There is no problem of balancing the currency. The rouble does not exist in the foreign trade of the U.S.S.R., nor the dollar inside it. In spite of its undoubted success, this is not a method which the countries with at least some degree of free enterprise would wish to adopt. They are willing to tolerate a certain degree of instability which arises from conflicting private and group interests rather than subordinate them all to the State.

Though the State is not omnipotent in the free enterprise countries, it is not powerless. It has a limited but important influence on the circulation of credit and currency. It has direct and advisory influence on industry. Finally it has the power of re-distributing wealth by taxation.

The last twenty-five years, after the war, have seen an unparalleled increase in prosperity in all industrial countries. Compared with the pre-war years in which all countries played the game of 'beggar your neighbour' we have seen also a remarkable improvement in international relations, as shown by the undreamt-of expansion of international trade. In 1969 it went up by not less than 16 per cent, far exceeding the growth of the national economies. Nevertheless there is still far too much insecurity. One may ask whether this could not be reduced by better economic prognoses, backed by the authority of the national planners? The answer is that a credibility gap has developed, not because of any dishonesty of the governments, but by their weakness to control events. In 1967–9 the European economy was shaken by the devaluation of the pound and of the franc, and by the revaluation of the mark. In all three cases the governments strenuously denied any such intention, until it happened. These declarations were meant as self-fulfilling prophecies, but they proved to be weaker than the self-fulfilling expectations of the speculators.

Speculation is anticipation at the expense of others, and is one of the strongest destabilizing agents in social life. For this reason the first entries in the list below are concerned with possible ways of suppressing unethical speculation. This, like everything in a free society, is a matter for compromise. Foresight is an economic virtue, and shall we punish those who have better foresight than others? We cannot adopt the practice of the U.S.S.R., where speculation in foreign currency is a capital crime, punishable by death or deportation. In spite of all its built-in instabilities our partly free economy has not functioned too badly, and all improvements must come piecemeal.

118. *Demonetizing gold*

For many centuries gold was a powerful stabilizing agency. It could not be faked by the princes or their alchemists, at worst it could be alloyed with base metals, and the coins were sometimes circumcised by the money changers. But gradually the stabilizing action of gold turned almost into its opposite, if for no other reasons, because there was so little of it. The gold hoard of the United States is little more than one per cent of its annual G.N.P. and the total amount of gold in the world probably not more than 5–6 per cent of it. Hence it is now the small gold tail which at times wags the huge economic dog; an obviously unstable and nonsensical situation. The last attack of gold on the economy was fortunately repelled. In 1968, at the time of the economically completely unfounded dollar-crisis, a free market was established in gold, where the price of the ounce rose temporarily to $47, until, in 1969 it fell back to the official price of $35, with heavy losses to the speculators.

The economy of the industrial countries has now grown far beyond the stage at which the currencies can be based on gold. They will have to find another base which cannot be faked: production. On the other hand gold may still be useful in the developing countries which are now in the stage which Europe reached in the early nineteenth century, and where confidence in governments is still very weak. The great Western countries could pass on their gold hoards in the form of loans to Asian, African, and Latin American countries, to form the reserves of stable currencies. The obvious objection to this is that just in these countries gold had a strong corrupting influence. Too many of the dollars and pounds which came to them as loans found their way into the vaults of the Swiss banks, in the form of gold bars. But once the Western countries have stopped hoarding gold, the price of gold is not likely to rise. At present half of the gold production is sufficient for 'industrial' use, mostly jewellery. If the corrupt politicians will hoard dollars instead of gold, they will do only half the damage which they are doing today, when they not only divert the dollars from the use to which they are intended, but pull the dollar down.

119. *International auxiliary currencies*

The 'Eurodollar' has now been for years a valuable stabilizing agency in European trade, but the best approach to the 'Bancor', the 'paper gold' proposed by J. M. Keynes (in a document preliminary to the negotiations which later led to the establishment of the International Monetary Fund and the International Bank for Reconstruction and Development*) is the S.D.R., the 'Special Drawing Rights' of the International Monetary Fund. In 1970 $3·5 billion will be added to the reserves of the members of the I.M.F., and $3 billion in the two succeeding years. Large as these sums appear, their total is only about a quarter of the world's monetary reserves in gold, but it is a promising beginning. It is a currency based on confidence in the strong economies of the Western countries.

120. *An international currency*

S.D.R. represents money circulating only among the Central Banks; it is not yet a currency in the usual sense. It appears by no means impossible to establish such a currency, without complete renunciation by the member nations of financial independence. Prior to 1914 the Latin Monetary Union between France, Belgium, Switzerland, and Italy performed very satisfactorily, and so has the Sterling Area since 1932. The European Payments Union, established as a condition of Marshall Aid, was invaluable in the reconstruction of the European economies and it is perhaps regrettable that it was later dissolved. It must be admitted though that monetary unions were much easier at a time when taxes amounted only to a small tip to the state, while today the budgets amount to 25–40 per cent of the G.N.P.s. Assume that states *A* and *B* have monetary union. *A* decides on a great plan for roads, new towns, schools, etc. *B* can rightly object that a good part of the materials, the food and clothing, etc. of the workers comes out of the production of *B*'s industries. The partners will probably not be satisfied until the state credits for capital accounts will balance, at least approximately, and retain similar ratios to the consumer good productions. This will require friendly agreement, and an acceptance of the principle that the standard of living shall rise parallel in the partner countries. Considering that the six countries in the European Common Market have accepted this principle, it is perhaps not too much to hope that all European countries will come to a similar agreement, and realize a common currency, promised by 1980. This would be the most radical way of stopping the dangerous flooding to-and-fro of 'hot money'. After the events of 1968 France lost about one-half of its reserves to Germany, and after the German

* *Proposal for an International Clearing Union*, Cmd. 6437, HMSO, April 1943.

revaluation a similar amount left Germany, though it did not all find its way back to France. The speculators had realized their ill-gotten gains, at the expense of almost everybody else. All states would gain by stopping this unethical practice, even at the cost of a fraction of their financial sovereignty.

121. *Stopping the 'free for all' in wage claims*

The 'revolution of rising expectations' has a rational basis: modern technology can indeed raise the standard of living up to the point of 'psychological saturation', which is not yet a problem of today. But the way in which it manifests itself is inflationary. Each trade union presents its claims separately. They have a feeling of injustice when the wages or salaries in one branch have dropped behind the rest, or even behind the price index, and they demand always more, to 'leap-frog' the others. The Prices and Incomes policy of Britain has completely broken down. In other countries even wages tied to the cost-of-living index were not sufficient to prevent costly industrial disputes, which usually end with a surrender to the workers of the branch concerned, at the cost of all the others.

In Britain all the efforts of the government have failed towards a *central* regulation of wage claims and towards making wage-agreements legally enforceable. To a great part this was due to the Trade Union Congress not having power over the branch unions, the branches over the local unions, and the local unions over the shop stewards. On the other hand the shop stewards have often far too much power over the workers, whom they terrorize. The obvious solution seems to be a permanent trade union congress, in which the branches of industry could (a) thresh out their conflicting claims between themselves, (b) present their decisions for acceptance, not to votes *per proxy*, but to the direct secret vote of the workers, and (c) present their wage-claims *as a whole* to the Government and to the employers, instead of in the usual 'round', that is to say in an inflationary spiral. This has been often proposed, and some countries have taken tentative steps towards it, but real progress in this direction has remained a pipe dream. Is it really impossible to solve this problem without exercising the power of the totalitarian state?

122. *Taxation*

In a modern industrial state, with a high standard of living, public expenditure will unavoidably increase compared with expenditure in the private sector. More roads, hospitals, schools, subsidized houses, etc. have to be built, which cannot pay for themselves, and relatively few factories, hotels etc., which can show a profit. The ratio will be even worse when the 'quality of life' will be taken seriously. Education alone may well require 15–20 per cent of the G.N.P. Economically, that is to say in terms of production of goods, there is no difficulty. A minority can very well provide goods for all. As a glaring example, in the United States the total number of agricultural workers, now about 6 per cent of the labour force, is expected to drop to 2·5 per cent by 1982. (*The American Economy*, McGraw-Hill, 1969.) But raising even a significant fraction of the costs of the welfare state by direct taxation is now out of the question. In France, where the total taxation amounts to 46 per cent of the G.N.P., only 18 per cent is raised by direct taxation, 43 per cent by taxes on expenditure, 39 per cent in social security contributions. Even in Britain, where direct taxation has long reached the stage at which, as J. M. Keynes said 'the avoidance of taxes is the only intellectual pursuit that still carries any award' and where private households save now only 6 per cent of their income, direct taxation brings in only 40 per cent of the public revenue, which is 38 per cent of the G.N.P.

In this situation, which is certain to worsen, the worried ministers of the future can do nothing else but to approach even further the practice of the communist countries, where practically all public expenditure is paid by indirect taxation. The example of France shows that this can be done without entirely discouraging the energies which display themselves in private enterprise.

123. *Reform of foreign aid*

In the last year, chiefly owing to the negative balance of payments of the United States, aid for the developing countries has dropped to a very low level. The foreign aid to India is now about two dollars per head per year. On the other hand the Russian aid to Cuba is about $20–30 per annum per head, and this appears just about sufficient to make a hard-working country, not at all badly favoured by nature, economically independent in 5–10 years. In order to become effective, foreign aid to the developing countries ought to be increased by about an order of magnitude for at least one or two decades, backed of course by a great educational plan. Economically, that is to say in terms of real goods and services, this does not appear impossible or even difficult. Now that the net foreign aid of the United States has been dropped to $2 billion, it can be said that the American worker works little more than *one* minute a day for the developing countries, while he is working about fifteen minutes for the Vietnam war. The difficulties are financial. The misdirected part of the foreign aid, instead of flowing into the right channels, finds its way into the Swiss banks, and weakens the dollar. As it is a public expenditure, it also adds to the burden of taxation. Is it really necessary to wait for an explosion, such as can be expected in Latin America, before we do something on an adequate scale?

4.4 Internal and international peace

Since the last war both international suspicion and international cooperation have developed at such an unprecedented rate, that it is impossible to estimate the balance of these two contrary forces. On the one hand the two superpowers, the U.S.A. and the U.S.S.R., have built up such stockpiles of nuclear explosives that they could wipe out all life on Earth, and such means of delivering them as could kill a good half of both their populations in a few hours. On the other hand, the old enemies France and Germany are peacefully cooperating in the E.E.C., and there were, in 1969, not less than 1600 international, non-governmental organizations, growing at 10 per cent per annum, in many of which U.S. and U.S.S.R. representatives worked together, as one team. World trade has increased by an order of magnitude since pre-war years.

Some tensions have decreased, but others have increased. Troops with loaded weapons are no longer called up against strikers in any of the industrialized countries, but they had to be called up in racial riots in the United States. The gap between the industrialized third of the world and the rest has increased. The ratio of the G.N.P.s *per capita* is now about 12 to 1 and is still growing. The population explosion is worst in the underdeveloped countries, and there can be no question now of solving this by emigration. It is hard to believe that as late as 1914 any Chinese or Indian could settle in the United Kingdom, without a passport, without any bothering by the police. The racial frontiers are now almost hermetically closed, and it is likely that they will be defended even at the cost of a nuclear war.

In the next fifty years we shall have to round a very dangerous corner. Some of the underdeveloped countries might reach the point of despair. In the highly developed rich countries a malaise is developing, of which the student riots are probably the first symptoms. These have decayed at the time of writing, but they are likely to return again, because there is a driving force behind them; the 'student explosion'. Nervous tensions inside the states threaten two dangers. Politicians, at the end of their tether, may turn towards the police state, or they might release the dogs of war.

There is now an interesting debate going on among psychologists and ethologists whether wars originate from an aggressive instinct, innate in human nature, or not. This is not entirely an academic question. Who could doubt that we should be safer if the heads of the U.S. and of the U.S.S.R. were men like Albert Schweizer or William Penn? Prehaps we could really solve the question of war with pacifying drugs, such as Arthur Koestler has proposed? But it will be better if the social engineer directs his attention to the powerful *amplifying mechanism* in our world. Its two driving forces are fear and self-interest. Self-interest can be allied to the will for power and domination, but it is dangerous enough if it takes on only the socially tolerated form of a drive towards *empire-building*; the enlargement of the establishment. It was no will towards international aggression, but only fear and empire-building which has made the U.S. army grow from its pre-war size of 90 000 to $2\frac{1}{2}$ million. (In 1967 $3\frac{1}{2}$ million.) The instrument in the United States is the 'military–industrial complex' (Eisenhower, 17 January 1961. Senator Fulbright in 1967 added the academics, and George Wald in 1969 the labour unions as the noble thirds and fourths to this complex.) We do not know exactly its counterpart in the U.S.S.R., but fear and empire-building by the military must be at work there too. With such powerful amplifying mechanisms there is no need for much aggressive drive from the top. Every engineer knows that if an amplifier is strong enough, any small amount of noise will start it off.

In trying to break up these immensely complicated, interwoven problems into separate items, I must emphasize even more than before, that even where I am hinting at possible solutions, I am chiefly concerned with bringing them to the attention of thinking minds.

124. *Breaking the cult of violence in the mass media*
Individual aggression may not be of decisive importance in modern war, but it is certainly important in internal peace. The question whether the appalling scenes of violence which are presented every

day on television screens, especially in the U.S.A., are contributors to violent crime and to riots, is one in which surveys have been very reticent in coming to definite conclusions. They appear to agree on the point that violence has a cathartic effect on the majority, but a violence-stimulating effect on a small minority. How these two effects balance is a moot point. Besides, TV is at most *one* of the factors which tend to make, in Rap Brown's words, 'violence as American as cherry-pie'. (There are also the gun laws, the films, and racial segregation.) Nevertheless, one cannot help suspecting that the cult of violence in American TV, compared with the much milder British programmes may contribute its share in making the ratio of murders *per capita* in the two countries 25 to 1. (This ratio would be probably even larger if one deducted in both countries the insanes and compulsives. According to Anthony Storr, *Human Aggression*, Allen Lane, London, 1968, in Britain one murderer in three attempts or commits suicide.)

125. *Student violence*

Student riots depend chiefly on two factors. One is the ratio of university students in an age group, the other is the permissiveness of the university staff and of the country. According to the Hudson Inst. Report No. 4, 1969, the number of university students in the U.S.S.R. was not much less than in the U.S. (3·6 million against 5 million in 1964, probably 5 million against 6·5 in 1969), yet there were serious riots in the U.S. and none in the U.S.S.R. From this, and also from the fact that in the States the more determined stand of the universities has strongly diminished student violence in 1969 compared with 1968, one must conclude that at the present stage the revolutionary movements of students are still fairly easily suppressed. But it would be most unwise to rely on this, until one has gone to the root of the trouble. This is that universities are traditionally institutions for hard-working and gifted students, who in the past, so long as their numbers were small, were rewarded with well-paid and socially distinguished jobs. This is no longer possible when, as in the States, more than half of an age group enters some sort of college. Why should the less-gifted work hard, when they could get the same sort of job without going to the university? I cannot see any way of making them work hard, nor any reason for it. I can see a solution only in a new type of teaching, which will be rewarding also for the lesser gifted, by being half entertainment, half instruction.

126. *The war of the classes*

When the socialist movement started, in the early nineteenth century, it took its strength from the masses of manual workers in the factories. (Farm workers were even more numerous at that time, but they could not be easily organized.) Meanwhile the proportions have greatly changed, manual workers are now a one-third minority in the United States and they will soon be a minority also in the European countries. But though class warfare has lost in ferocity, it has not lost in strength. An underpaid clerk can hate his boss just as much as a lathe operator, and teachers can strike just as well as factory workers. The practical criterion is not one of social justice (who can measure that?), but who can drop a spanner into the social machinery. When, as is quite likely, the number of manual workers will be reduced to less than 20 per cent of the labour force, they will be able to exert just as much pressure on the rest as when they were 80 per cent.

The best solutions of the class warfare problem have so far been produced by two very different countries, Japan and Sweden. In Japan it is the unwavering traditional loyalty of the workers to the firm, to which they are wedded for life, in Sweden it is the cool-headed, reasonable approach of the trade unions, who decide in discussions with the employers how much they can take out of the pot without ruining the firm. There is of course also a Russian solution. They have abolished class warfare without abolishing classes, simply by making strikes illegal.

In the free countries a vestige of class warfare is likely to remain permanently, because industrial disputes give excitement, a feeling of unity for groups, and add to the spice of an otherwise too monotonous life. But a better solution has been proposed by Karl Marx himself (following Fourier). A man shall have several occupations in life. He can take a turn as a manual worker, a longer one as a service worker, and why not also, if he is willing to face the hard work and the responsibility, a turn in the administration? I would like to add also, work in different countries for those who do not mind the trouble of learning a foreign language.

127. *Swords into ploughshares*

Elsewhere I have emphasized that a high civilization requires a *sedentary* population, one which takes roots (which does not of course exclude frequent travel, or even years spent in foreign countries), and that this is hardly possible where the local factories can shut down from one day to the other. But armaments factories *ought* to shut down. It is not surprising that labour protests against being thrown out of a job, even if it is a job of making machines for killing. ('Morals Sir?' asks Shaw's Mr. Doolittle, 'I can't afford 'em.') It is the same, but rather worse, with the technologists who are now employed in the armaments research establishments.

It is not impossible, but a rather long job, to convert an armaments factory into one for civilian products. Krupp's have done it on an enormous scale after the first World War. They were helped by the German trade union structure; the workers themselves could rebuild the factory. This is not possible under the British or American system, where many workers are organized by crafts. A reform of this system, at least in the armaments industries, would certainly not be easy, but it would help in the solution of an extremely difficult problem.

As regards the 'scientists' who are now working on new weapons, among them on nerve gases and exceedingly violent viruses, I wish that they could be ostracized by the scientific communities in the countries that have not yet followed President Nixon's courageous unilateral renunciation of biological warfare. But there may be a less radical way. For instance, as has been proposed, the microbiological establishment at Porton, England, with its wonderful equipment, could be converted into a laboratory for producing enzymes and other difficult organochemical substances.

128. *Breaking the feedback loop of fear*

The military-industrial complex in the U.S.A. and its counterpart in the U.S.S.R. are certainly overgrown, but who can say by what factor? They are fed by mutual fear. Any reported increase of missiles or of Polaris-type submarines, or A.B.M.s on the one side must be balanced on the other side. One must admit that the fear on the U.S. side is not unfounded. We know that nobody could lay his head into Stalin's lap, and at any rate since the occupation of Czechoslovakia we know that it would be risky to try this with Brezhnev's. But Brezhnev's move was also caused by fear: by the fear of Czechoslovakia becoming independent on the pattern of Tito's Yugoslavia, and other satellite countries following the example, until another hundred million or so people were added to Russia's potential enemies. The error in this rational calculation is only that the occupation of Czechoslovakia has not added a tittle to Russia's safety against nuclear attack. On the contrary, it has strengthened the worst suspicions.

How could we break this terrifying feedback loop of fear? At the time of writing, negotiations towards a slowing down of the armaments race are going on. If, as is not unlikely, they are delayed or end only with a *pro forma* agreement, we can gain hope from the fact that on our side a certain peace-strategy is already gaining acceptance: if there is a report on a new weapon-development on the other side, *do not try to leap-frog them.* Delay the countermeasures to the last possible moment. This gives a wide latitude, of several years, because the nature of nuclear warfare is such that even a first-strike superiority of 10 to 1 is not sufficient to save the aggressor from the counterstrike, the 'dead man's revenge', in which something like one-half of the population of the aggressor and almost all his cities would be destroyed. Such a policy, consequently followed for many years, would go a long way to allay the fears of the U.S.S.R., now often expressed in their military journals, that 'according to reliable reports the U.S. are preparing for a devastating attack'.

129. *International citizens*

From these measures, on which decisions can be taken only at the highest levels, let us now descend to more modest ones, which are within the reach of the ordinary citizen. I have already emphasized (in my criticism of translating machines, No. 58), how important I consider the growth in the number of bilingual or multilingual people, who could become citizens of the world. I take some further suggestions from the eminent Norwegian peace researcher, Johan Galtung. (On the Future of International Systems, *Mankind* 2000, Oslo, Universitatsforlaget, London, Allen & Unwin, 1969.) He predicts the rapid emergence of *international peace specialists* and of *international development engineers.* He also predicts

'a steady growth in the mutual interpretation of all developed, industrialized nations with neomodern segments with each other, using INGOs (international non-governmental organizations) and IGOs (international governmental organizations) as building structures and individuals with cross-, trans-, and supra-loyalties as building blocks'.

Let us hope that these ideas will spread among our young people, who certainly do not lack idealism, but are often in need of suggestions for constructive outlets for their energies.

4.5 Towards a stable, mature society

Quoting from myself (*Nature*, London, 1963):

'Exponential curves grow to infinity only in mathematics. In the physical world they either turn round and saturate, or they break down catastrophically. It is our duty as thinking men to do our best towards a gentle saturation, instead of sustaining the exponential growth, though this faces us with unfamiliar and distasteful problems.'

In our technological civilization *growth* has become synonymous with *hope*, and man cannot live without hope. The *ananke* of the Greeks, the pressure of a stinting and hostile nature, could not have been defeated without science and technology. Man, used to thousands of years of scarcity, has expanded into the living space created by technology, and it would be blind optimism to believe that this growth was all organic, evolutionary. There is a pure quantitative element in it, which one can call cancerous. Unless we separate evolution from multiplication, quantity from quality of life, the prospects for our civilization are dim.

The first condition for this is arresting the growth of population. On the whole globe this cannot be achieved from one year to the other, it is rather a matter of a hundred years, during which the developing countries are likely to suffer hardship. The prospect is better in the industrial countries, and the example of Japan has shown that an (almost) stationary population need not arrest dynamic industrial growth. This growth too will have to slow down and stop some time, simply because the material needs of men are not infinite. But it must not stop because of the *nausea*, which has already visibly infected a part of the young people in the industrially most successful countries. It must stop by the transition into a new stage of civilization, which offers *hope without material growth*.

How our time needs prophetic poets who could give hope to the young people! Their thirst is so great that they have made the almost unreadable philosopher Herbert Marcuse their prophet, together with the tyrant Mao and the romantic revolutionary Che Guevara. Unfortunately, our poets and writers are too busy expressing disgust and despair, presenting our world as even more meaningless than it is. So it falls to the thinkers to carry the heavy burden. But the cold voice of reason need not be powerless if it is allied with imagination.

The only at least partly optimistic utopia written in the last forty-five years outside Russia has been Aldous Huxley's *Island*, and even his happy little island is destroyed in the end by the adolescent nationalism, with which we are only too familiar. And yet, utopianism is not dead, it has found its last refuge in the promises of politicians, which are at most only half believed. But politicians and their economic advisers are increasingly entangled in day-to-day problems. It is necessary to mobilize a force of thinkers, such as the *Encyclopédistes* were in the eighteenth century, to provide visions thirty or more years ahead. There are some promising signs that such a movement is on the way. This will be the subject of the first entry.

130. *Futuristic studies*

Some writers, such as Bertrand de Jouvenel, have objected to the term 'futurology', because an 'ology' is a science which deals with things that exist, and the future *per definitionem* never exists, so let us call it 'futuristics'. Three such groups have distinguished themselves in the past, the 'Futuribles' group in France, around Bertrand de Jouvenel, which constructs 'possible futures', the group around Olaf Helmer, formerly in the RAND Corporation, now in the Institute for the Future, Middletown, Conn., which collects DELPHI forecasts, and is now developing more advanced methods, and the Hudson Institute, under Herman Kahn, which constructs general avenues of future developments, and also 'scenarios'; detailed contingent forecasts, which cannot be properly called utopias because most of them are undesirable. There are in addition also privately subsidized 'lookout' organizations, such as TEMPO and Systems Development Inc., both in California. Computer models of society have been developed by Jay Forrester in the MIT, Abt Associates, Cambridge, Mass., and Prof. Richard Stone, Cambridge, England.

All these groups are well aware that the exponential growth cannot continue indefinitely, though some of them also supply their industrial customers with forecasts for shorter times, which are almost exponential. The development of computers is gradually enabling them to deal with extremely complicated models, so long as the factors are quantifiable. The weakness of all such forecasts is of course that one does not know how human nature will react to new circumstances. It has been often suggested that these could be obtained from social experiments; 'probes'. Their weakness is of course that it is usually cranks who volunteer for such experiments, not random samples of humanity. Evidently, forecasting the future can never be an exact science. Nevertheless, these institutions, supported by developments in practical psychology and an infusion of an imaginative *normative* element, are still our best hope that the coming revolution will not find us quite unprepared.

131. *Slowing down the consumer society*

The present trends in the U.S., as estimated by the Economics Department of the McGraw-Hill Publishing Company, are well expressed in the accompanying figures:

Predicted consumption in the United States, 1967–82

	Year	1967	1972	1977	1982
	1967 dollars	491·6	603·0	745·5	930 billion
Consumer spending	%	100	122·2	151	188 %
	1967 dollars	72·1	90·5	115	147·5 billion
Durables	%	100	125	159	204 %
	1967 dollars	217·5	261	316·5	387 billion
Nondurables	%	100	120	145	173 %
	1967 dollars	202·0	251·5	314·0	395·5 billion
Services	%	100	124	155	196 %

On the whole this shows fairly uniform growth, at an annual rate of 4·2 per cent, with consumer's durables still leading. This, one can presume, is what businessmen expect and hope for. The growth of the U.S. population during these fifteen years is estimated as from 199·1 million to 242·4 million, or by 22 per cent, that is to say the increase in consumer spending *per capita* is expected to be 52 per cent. This may not appear unreasonable, but let us step just another ten years ahead, to 1992, and we get a doubling of expenditure per head. What does this mean? Four cars in the garage instead of two, and four colour television sets per family? If anybody still considers this as reasonable, let him add on another ten years, or another twenty. The world, let us hope, will not come to an end by that time, but consumer expenditure *must* have approached some sort of ceiling. If by that time we shall have any sort of reasonable world, the spending on services must far exceed the other two items, which ought to reach saturation well before the end of the century, even if every scrap of poverty will be eliminated in the United States.

The U.S. statistics are of particular interest, because this country has just passed the threshold to the 'post-industrial society', with a G.N.P. *per capita* of $4000 and almost $10 000 spending money per family. At the past rate of growth the European industrial countries are 10–25 years from this level. Japan is a special case, because by simple trend extrapolation the G.N.P. *per capita* ought to exceed that of the U.S. before the end of the century. On the other hand many of the underdeveloped countries are not likely to reach by that time what in the U.S. is called the 'poverty level'.

Material growth must continue in the developing countries, with as much help as the rich countries can give them, for many decades, but it must be transformed into qualitative growth in the happier third or quarter of the world. The first problem is how to convince the dynamic individuals in our society that this is not the end of the world. Much of the talent and of the energy in our world is concentrated in the managerial body of the great firms. The rising production chart behind the executive's desk is the symbol of his supreme value: growth. It is these people who need to be convinced that the great transformation will be just as much in need of their energies as the present system. Of course, the amount of goodwill, smiles, love, and happiness in the world is not as easily measured as the output of motor cars.

132. *Hope in an economically stationary society*

Man could never live without hope, but hope has become synonymous with growth only in the technological society. The manager or the professional can climb up the ladder in his profession even in a stationary society. But what other hope is there for the workman, who enters a factory at sixteen and leaves it at sixty, doing the same job, unless by the general rise in the standard of living, or by the fight of his trade union he can increase his wages? Wages rising with age are certainly a factor in the great stability of the Japanese society. It is regrettable that this good practice is now gradually being abandoned in Japan, in favour of efficiency. A stationary society can not only afford this principle; it can hardly exist without it. Hope is an individual value, its substitution by class- or collective values was one of the mistakes of our industrial society.

A change of occupation around the age of forty, and progressive life-long education could do even more for maintaining hope as a value and as a drive than rising wages in a mature society.

133. *The upgrading of service occupations*

The labour force employed in services is steadily rising in all industrial countries. As an example, in New York City between 1952 and 1969 it has risen from 520 000 to 750 000, while in the same time the labour force in the manufacturing industries has dropped from 1·05 million to 850 000. It must be suspected though that this has not substantially improved the quality of life in New York City.

If we really want to improve the quality of life, we must visualize a post-industrial society, in which a tenth, or perhaps even a quarter of the population are on vacations from the economic life, some on holiday, some on sabbatical education courses. It cannot be helped that in such a world the whole immense shore of the Mediterranean will change into a Torremolinos, from end to end. But if the holiday-makers will have to live in automated hotels, life will hardly be worth living. A leisured, civilized world requires smiling personal service, but not of course from a *class* of servants. Everybody, except perhaps professionals such as medical doctors (who are really also service workers), ought to learn and practise for a span in his life some sort of personal service. We must wipe out the stigma which from the time of slavery still attaches to such occupations as shop assistants, waiters, cooks, and domestic service workers.

My friend J. F. Engelberger has made an interesting suggestion: put them in uniform, like the English commissionaires, the night watchmen, or—more glamorous than all—air hostesses. Give them a special education and an *esprit de corps*. I consider this as a model of a social invention; one which makes use of human weakness for a good cause.

134. *New education: the U.K. Open University*

This is an interesting experiment, probably the germ of even more important developments. Its purpose is to give the equivalent of university degrees to those people who 'did not have a chance', but there is nothing to prevent it from being extended to the greater number of those who had a chance but did not use it. Each student gets tuition in four ways; by correspondence, by radio or television, by summer schools (2 weeks per year) and by local tuition. It will require about four years for a student willing to work about ten hours a week to acquire an honours degree. There is a fee (£140 for a degree, £180 for an honours degree), small enough for any worker to afford it, large enough not to be abandoned lightly. Each person has to take two Foundation Courses, to be chosen from Mathematics, Science, Understanding Society, and Humanities. About 25 000 students are expected in the first year.

135. *New education: the mass university*

In the opinion of experienced educators, the traditional university education 'takes' on perhaps 5 per cent or at most 12 per cent of the population. These are the ones with IQs of 120–125, provided that they have also the right amount of motivation. This is also more or less the proportion of well-paid and socially distinguished jobs which an industrial society can offer. (The suggestion, put out by some authors, that the 'post-economic' society may be developing towards 100 per cent 'quaternary' occupations I consider as unmitigated nonsense.) But the universities in France take already about 25 per cent, the colleges in the United States about 50 per cent of an age group. No wonder that a survey carried out in the United States (by the Yankelovich Organization, for *Fortune* magazine, October 1968) showed that only 58 per cent of the students expected to earn more money and have more interesting careers by studying, while for the rest 'college means...perhaps the opportunity to change things rather than make out well in the existing system'.

Though most of these young rebels will be probably broken in when they marry and become responsible for their families, we cannot be happy about a system in which the university fails in its main function: imparting knowledge and the love of knowledge. A new education must be created, a mixture of entertainment and instruction, which makes the less gifted appreciate the complicated civilization in which they live. The technical means are now available: music, films, travel, games; it is up to the educators to make efficient tools of them.

136. *Lifelong education*

It is also the experience of educators that mature people, past forty, in evening classes, are on the whole more appreciative and are more rewarding to teach than the average of young students. These people, who sacrifice their evenings, are of course a selection. For the majority it will be necessary to have sabbatical weeks, months, or even years. Knowledge increasing with years is now the privilege of a small class of people who have never stopped learning; this privilege must be extended to all. It will be of course expensive in the terms of an economics of scarcity, but a vital necessity in the super-affluent society of tomorrow. Teaching, after the age of forty, is one of the service occupations for which suitable gifted people can be trained or retrained, and something to look forward to in the years when they will work in production jobs.

137. The 'Moral equivalent of war'

Have we made any progress towards this problem since William James first posed it in 1905? In a sad way, we have. Modern war is no 'moral' equivalent of war as it still was in William James's time. Releasing nuclear bombs, or canisters of bacteria over an enemy town gives little satisfaction to the aggressive instincts of man, and releasing the catches and pressing the buttons for the launching of an I.C.B.M. gives even less. I have written elsewhere that the feelings of a fighter pilot in the Battle of Britain may not have been very different from those of a cavalryman at Omdurman, but since that time we have taken a further fateful step towards the dehumanization of war.

It is quite likely that a psychological equivalent of war has spontaneously developed in our society; but one which certainly cannot be called 'moral'. In the twenty-five years since the last war, which were an epoch of an unparalleled rise of prosperity in the industrial countries, crimes of violence have grown at a frightening rate. At the time of writing it had just been announced that in 1969 crimes of violence in London increased by 29 per cent. Lawlessness is now approaching the level which it may have had in the frontier days, in the Wild West, in spite of a police organization which was undreamt of in the bad old times.

A great part of this is not *individual* violence; it is fed by the gang spirit which is a degraded form of the comradeship, the platoon-spirit, in an army. The gang, the cell is an immensely strong unit. It may be easier to convert into a shock-troop for some good purpose than to break it up into individuals. It must be admitted, though, that such 'gangs for the good' would lose their strongest cohesive force: the disapproval of society.

All utopian thinkers, from Plato to Wells and Yefremov, saw at least a part of the solution in forming out of the most energetic individuals an elite body (Guardians, Samurai, Herculean feats), whose education contains an element of *hardship*. It is very likely that our permissive society will have to recognize this as a healthy principle if it wants to survive. The more permissive a society, the less can it do without a hard apprenticeship.

In this little book I have not set out to solve the problems of the world, but to bring at least some of them to the attention of thinking people, especially of the younger generation. Let me hope that I have made at least some of them realize that taken together these problems constitute a challenge more exciting and far more rewarding than the conquest of space.

Index

Index